R00602 62060

REF

QE 369 .06 E37 1987

Ehlers, Ernest G.

Optical mineralogy

$39.95

DATE			

BUSINESS/SCIENCE/TECHNOLOGY DIVISION

© THE BAKER & TAYLOR CO.

Optical Mineralogy
Theory and Technique

Volume 1

Optical Mineralogy

Theory and Technique

Ernest G. Ehlers
Professor Emeritus
Ohio State University

BLACKWELL SCIENTIFIC PUBLICATIONS
Palo Alto Oxford London Edinburgh Boston Melbourne

Editorial Offices

667 Lytton Avenue, Palo Alto, California 94301
Osney Mead, Oxford, OX2 0EL, UK
8 John Street, London WC 1N 2ES, UK
23 Ainslie Place, Edinburgh, EH3 6AJ, UK
52 Beacon Street, Boston, Massachusetts 02108
107 Barry Street, Carlton, Victoria 3053, Australia

Distributors

USA and Canada
Blackwell Scientific Publications
P.O. Box 50009
Palo Alto, California 94303
(415) 965-4081

Australia
Blackwell Scientific Publications (Australia) Pty Ltd
107 Barry Street, Carlton
Victoria 3053

United Kingdom
Blackwell Scientific Publications
Osney Mead
Oxford OX2 0EL

Sponsoring Editor: John H. Staples

Production: Mary Forkner, Publication Alternatives

Text and Cover Design: Gary Head

Manuscript Editor: Andrew Alden

Illustrations: Robin Mouat

Compositor: G&S Typesetters

© 1987 Blackwell Scientific Publications

All rights reserved. No part of this publication may be reproduced, stored in a retrieval system, or transmitted, in any form or by any means, electronic, mechanical, photocopying, recording or otherwise without the prior permission of the copyright owner.

Library of Congress Cataloging-in-Publication Data

Ehlers, Ernest G.
  Optical mineralogy
  Bibliography; p.
  Includes index.
  Contents: v. 1. Theory and technique.
  1. Optical mineralogy. I. Title.
QE369.06E37   1987   549'.125   86-14793
ISBN 0-86542-323-7 (v. 1)

British Library Cataloguing in Publication Data

Ehlers, Ernest G.
  Optical mineralogy.
  Vol. 1: Theory and Technique
  1. Optical mineralogy 2. Microscope and microscopy
  I. Title
  549'.125   QE369.06
  ISBN 0-86542-323-7

For Diane

Contents

Color Plates:

Plate 1. Michel-Lévy Color Chart
Plate 2. Interference Figures and Sign Reactions (see Chapter 6)

Preface xiii

1 The Polarizing Microscope 1

Parts of the Microscope 1

Care of the Microscope 4

Photomicrography 4

2 Light and Matter 6

Wavelength and Color 6

Frequency, Energy, and Velocity 8

Plane-Polarized Light 8

Light Transmission through Isotropic and Anisotropic Materials 10
   Color 10
   Atomic Polarization by Light 11

Interference 13

3 Examination of Isotropic Substances 16

Index of Refraction 16

Determination of Index of Refraction 18
- *Relief 19*
- *The Becke Line 21*
- *The Double Becke Line 22*
- *The Oblique Illumination Method 24*
- *Determining n in Thin Sections 24*

Identification of Isotropic Materials 25

4 Uniaxial Materials and Light, I 26

Uniaxial Materials and Polarization 26

Ray Velocity Surfaces 29

The Optical Indicatrix 32

Uniaxial Crystals Viewed with a Microscope 33
- *Optic Axis Parallel to the Stage 34*
- *Optic Axis Inclined to the Stage 35*
- *Optic Axis Perpendicular to the Stage 36*

Measuring Indices of Refraction of Uniaxial Materials 36

5 Uniaxial Materials and Light, II 40

Cleavage and Fragment Orientation 40

Interference Effects with Monochromatic Light 44
- *Full-Wave Retardation 45*
- *Half-Wave Retardation 48*
- *Quarter-Wave and Intermediate Retardations 49*

Interference Effects with Polychromatic Light 51

Accessory Plates and Wedges 53

Using Accessories to Determine Vibration Directions in Immersion Mounts 54

Using Accessories to Determine Vibration Directions in Thin Section 56

Pleochroism 56

6 Uniaxial Interference Figures 58

Uniaxial Optic Axis Figures 59

Uniaxial Inclined Optic Axis Figures 62

Uniaxial Optic Normal Figures 65

Uses of Uniaxial Interference Figures 68
- *Determination of Anisotropic Character 68*
- *Crystal Orientation and Index Measurement 68*
- *Determination of Sign 69*

7 Microscopic Identification of Unknown Uniaxial Materials 75

Oil-Immersion Technique 75

Thin-Section Technique 77

Additional Factors 79
- *Twins 79*
- *Anomalous Interference Colors 79*
- *Variation in Index of Refraction 79*

8 Biaxial Materials and Light 81

Indices of Refraction and the Indicatrix 82

Vibration Directions in Biaxial Crystals 85
- *Two Principal Planes Perpendicular to the Stage 85*
- *One Principal Plane Perpendicular to the Stage 89*
- *No Principal Plane Perpendicular to the Stage 90*

Biaxial Interference Figures 91
- *Acute Bisectrix (Bxa) Interference Figure 91*
- *Obtuse Bisectrix (Bxo) Interference Figure 100*
- *Optic Normal (O.N.) Figure 101*
- *Interference Colors and Figure Type 101*
- *Distinguishing among Centered Bxa, Bxo, and O.N. Figures 101*

Estimation of $2V$ 103

Off-center Interference Figures 104

Determination of Optic Sign 108
    Acute Bisectrix Interference Figures 108
    Obtuse Bisectrix Interference Figures 109
    Optic Normal Interference Figures 110
    Optic Axis Interference Figures 110

Common Problems 111

9 Optical Behavior of Biaxial Materials as Related to Crystal Symmetry 114

Optical Orientation of Orthorhombic Materials 114
    Dispersion 116
    Cleavage 117

Optical Orientation of Monoclinic Materials 119
    Dispersion 119
    Cleavage 123

Optical Orientation of Triclinic Materials 123

10 Microscopic Identification of Unknown Biaxial Materials 126

Identification with Oil-Immersion Techniques 127
    Interpretation of Published Data 127
    Interpretation of Data from an Unknown Material 130

Identification in Thin Section 133
    Color and Pleochroism 133
    Interference Colors 134
    Birefringence 135
    Indices of Refraction 135
    Cleavage and Habit 135
    Elongation Direction 135
    Extinction Angles 135
    Other Factors 137

11 Special Techniques 138

The Detent Spindle Stage 138
    Setting Up 138

Interference Figures 139
Orthoscopic Observations 140

The Universal Stage 146
Description 146
Observations 148

Bibliography 151

Index 155

Preface

Optical Mineralogy provides in two volumes the necessary information to identify the common rock-forming minerals in both thin sections and grain mounts by use of the petrographic (polarizing) microscope.

Volume 1 describes optical theory and techniques. The use of grain mounts is emphasized in the explanations as their use facilitates an understanding of optical theory. The described optical techniques are useful, however, for both grain mounts and thin sections.

A prior knowledge of introductory crystallography and mineralogy is assumed. No background in introductory physics is assumed, as the behavior of light is discussed in Chapter 2. The order of presentation is arranged to be compatible with a standard undergraduate university course in optical crystallography; this has led, in a few cases, to the applications of a technique being covered before the underlying theory is fully discussed. The length of the book has precluded an exhaustive coverage of optical theory; extended discussions are found in Bloss (1961, 1981), Hartshorne and Stuart (1960, 1969), Johannsen (1918), Wahlstrom (1979), and Winchell and Winchell (1937). These and other sources are listed in the Bibliography.

The spindle stage, described in Chapter 11, is currently receiving limited but increasing usage at the undergraduate level. It is, however, an inexpensive device that can be attached to any petrographic microscope. In addition to obtaining fundamental optical data, it is extremely useful in showing the directional relationships among interference figures and crystallographic parameters.

In Volume 2, descriptions of more than 150 common rock-forming minerals are presented. The mineral descriptions often include both color and black-and-white photomicrographs of thin sections. A departure from standard practice is the alphabetic arrangement of mineral groups and species in both the descriptions and color plates. In addition to standard descriptions, information is furnished on indices of refraction, extinction

Ernst Leitz Wetzlar GmbH

1

The Polarizing Microscope

Don't stand shivering upon the bank; plunge in at once and have it over.
SAM SLICK

Solid materials can now be studied with highly sophisticated techniques, which include X-ray diffraction analysis, electron microprobe analysis, and scanning electron microscopy. With such approaches available, it may seem a waste of one's youth to learn how to use a petrographic microscope—an instrument that was invented about 150 years ago for the examination of transparent substances.

There are a number of reasons for the continued use of the petrographic microscope. One of these is phase identification. A skilled microscopist can identify nearly any solid transparent material in less than a half-hour (and usually in just a few minutes). If an aggregate of several materials is present, it is relatively easy to identify those substances that make up less than 5 percent of the whole; this feat is generally impossible with X-ray diffraction techniques. Indeed, it is often possible to identify a single sand-size fragment within a mixture. Atmospheric dust particles are commonly identified by microscopic analyses, as are the fibers that compose insulating and ceramic materials.

The second major use of the petrographic microscope is textural analysis. This is usually done with the aid of thin sections: cut and polished slices of material (usually 0.03 mm in thickness) that are glued onto glass microscope slides. Analyses of thin sections reveal the types of materials present, grain sizes and distribution, the compatibility or incompatibility of minerals, the distribution of inclusions within minerals, and the character and intensity of secondary reactions (such as chemical alteration and exsolution). The shapes and orientations of the various minerals may reveal their sequence of formation and may supply evidence for later periods of deformation or recrystallization. A quick glance at the thin section of a rock is usually sufficient to indicate whether its origin is igneous, sedimentary, or metamorphic.

The cost of an adequate petrographic microscope is modest—a few thousand dollars—as compared to more sophisticated instruments (electron microprobes cost around $500,000). Once in place in a laboratory, the operational cost of a microscope is slight.

The fundamental techniques in microscopy can be mastered within a quarter or semester course at most universities. Used in conjunction with other techniques, the petrographic microscope is a valuable tool for the geologist, chemist, solid-state physicist, pedologist, ceramic engineer, and metallurgist.

PARTS OF THE MICROSCOPE

Petrographic microscopes are made by a variety of manufacturers (including Wild, Olympus, Leitz, Nikon, Vickers, and Zeiss). Although dif-

2 Chapter 1: The Polarizing Microscope

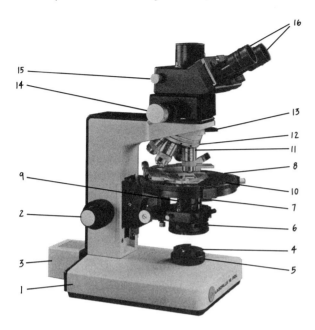

Figure 1-1
A modern petrographic microscope, the Laborlux 12 Pol (courtesy of E. Leitz, Inc.). Parts referred to in the text are (1) the stand, (2) coarse and fine vertical stage adjustment, (3) lamp housing, (4) filter holder, (5) field diaphragm, (6) lower polarizer, (7) aperture diaphragm, (8) stage, (9) condensing or converging lens, (10) vernier scale, (11) objective lenses, (12) nosepiece, (13) accessory slot, (14) upper polarizer, (15) Bertrand lens, and (16) oculars. The vertical tube at the top is used for mounting a camera. (Photo courtesy of J. Hinsch and E. Leitz, Inc.)

fering considerably in both design and quality, all have certain basic features in common as shown in Figure 1-1. Numbers in the figure correspond to those in the text.

The microscope stand (1) provides a rigid base for the entire assembly. The stand contains one or two knurled knobs (2), used to bring a viewed object into focus (that is, to bring the object into ideal viewing position). This is accomplished by varying the distance between the lens system and the specimen.

The source of illumination in older or less expensive instruments is an adjustable mirror with a separate light source. Most modern instruments use a permanently attached lamp (3) of fixed or variable intensity. At or directly above the light source, a holder (4) may be provided to accommodate various filters. Just above the light source there may be an iris (field) diaphragm (5), used to regulate the amount of light transmitted through the specimen. Its opening is adjusted with a rotatable lever arm or ring. In general, the amount of light used should be minimal (consistent with proper viewing) in order to reduce eyestrain during prolonged observations.

The light, after passing through the iris diaphragm, goes through a polarizer (6), located in the substage assembly. The polarizer in most modern instruments consists of a sheet of Polaroid, which converts normal light into plane-polarized light (discussed in Chapter 2). The first polarizers (made of calcite) were invented by William Nicol in 1828; the term "nicol" continues to denote a variety of polarizing devices currently used in microscopes. The polarizer (6) in the substage assembly is called the lower polarizer (lower polar, or lower nicol), as another one is located higher in the microscope tube.

Light from the lower polarizer, after passing through a second iris diaphragm (7) called the aperture diaphragm, may go directly upward through the sample, which is placed on the microscope stage (8), or it may first pass through a converging or condensing lens (9). The condensing lens can be rotated into or out of the optical system.

The microscope stage, upon which microscope slides are placed, has a central opening that allows light rays to pass through the sample. The stage rotates freely, and is calibrated with a vernier scale (10) to permit measurement of degrees of rotation. A pair of stage clips may be available to hold the sample slide firmly on the stage, or a mechanical stage may be used to manipulate the sample precisely.

One or more compound lenses are attached to the base of the microscope tube, directly above the sample. These are called objective lenses, or simply objectives (11). As various degrees of magnification are necessary, most petrographic microscopes use three objectives; they may be mounted individually or be attached together on a rotating device called a nosepiece (12). The objective lenses should be parfocal; that is, after focusing one lens, each of the others is also in focus when put into the viewing position. Also, each objective must be centered on the axis of rotation of the stage, as explained in Figure 1-2. Stamped or etched on the side of the objective are two numbers; the first is the initial magnification or power (such as 10×) and the second is the numerical aperture, a feature whose importance is described in Chapter 8.

Directly above the objective lens is a diagonally oriented opening in the microscope tube (13) that permits the entry of an accessory device: a quartz wedge, quarter-wave mica plate, or first-order red plate. The use of accessory devices is discussed in Chapter 5.

Above the accessory slot is the upper polarizer (14), also called the analyzer, upper nicol, or upper polar. This device is similar to the lower polarizer in the substage, but has two important differences: it can be easily inserted or removed from the optical system, and its polarization direction is normally oriented perpendicular to that of the lower polarizer. The next higher device usually present is the Bertrand lens (15), sometimes called the Amici-Bertrand lens. It, too, can be moved into or out of the optical system. When in place this lens is focused on the upper surface of the objective lens; the specimen cannot be seen. Instead, the operator can observe light interference images that are produced by most crystalline materials. Interference figures are introduced in Chapter 6.

The uppermost compound lens is called the ocular, eyepiece, or field lens (16); it further magnifies the image that has been produced within the microscope tube by the objective lens. Between the two lenses of the ocular are located the cross hairs. A slot in the microscope tube keeps

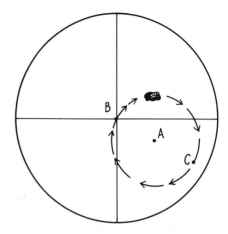

Figure 1-2
When the objective lens is centered, an object located at the cross-hair intersection remains in place during rotation of the stage. If the objective is off center, it must be adjusted using the centering wrenches furnished with the microscope. One procedure is to first place a recognizable object at the cross-hair intersection (B). Rotate the stage and note the apparent center of rotation (A). Move the objective with the wrenches until the apparent center of rotation is at the cross-hair intersection, then test the centering again and adjust as necessary. An alternate method is to rotate the stage until the object is at its farthest distance (C) from the cross-hair intersection, then bring it halfway back to the cross-hair intersection (point A) with the wrenches. Repeat the procedure until the objective is centered. Objectives of some microscopes are centered by turning rings on the objective; objectives of others are fixed, and the stage is centered instead.

the ocular and cross hairs fixed in proper orientation. The cross hairs are brought into focus by rotating one knurled ring on the ocular against another. A monocular microscope has a single ocular; the binocular microscope in the figure has two. The magnifying power of the ocular is stamped or etched on its side or top.

Assuming no additional lenses within the microscope, the total magnification is found by multiplying the magnification of the objective lens by that of the ocular. Thus, a 45× objective with a 10× ocular yields a total magnification of

450×. Many microscopes have an additional magnifying lens built into the tube; if present, this will be noted on the microscope.

When using a monocular microscope, eyestrain can be minimized by keeping both eyes open. Many wearers of eyeglasses can use the microscope without glasses, because any near- or farsightedness (spherical aberration) is easily compensated by moving the stage up or down. Astigmatism, however, is corrected with elliptical (toric) lenses; glasses for astigmatics distort the image as they are held facing a page and rotated. Astigmatics must wear glasses while using the microscope, but special high-focus oculars are available for those with glasses.

CARE OF THE MICROSCOPE

When not in use, the microscope should be protected by a plastic cover to keep it free from dust. Alternatively, return the microscope to its carrying case; be sure to lock the case to prevent the microscope from falling out.

A crisp image is obtained only when the glass surfaces of the microscope are clean. Dust particles on exposed lens surfaces can be removed by a jet of air, or by means of a sable or camel's-hair brush (that has been previously cleaned in ether). Oil and grease can be removed from lens surfaces with lens paper that has been moistened with either xylene or Zippo lighter fluid. Kodak lens cleaner applied with a cotten swab can also be used to clean microscope lenses.

For particularly resistant adhering particles, additional cleaning with styrofoam (expanded polystyrene) has been found to be quite effective. This type of white granular material is well-known as a packing or insulating material. Break off a piece, press it against a dry lens surface, and rotate the lens coaxially. Be sure that no xylene droplets are present, as xylene dissolves styrofoam. Any remaining grains of adhering styrofoam can be removed by blowing or with a clean sable brush.

In addition, always be certain that the glass slide or thin section that is on the microscope stage is clean, as this is part of the optical system. Ordinary tissue is sufficient for this purpose.

PHOTOMICROGRAPHY

Photomicrography is a subject that is usually learned when research-grade microscopes are available. An example of such a microscope is shown in Figure 1-1, where, in addition to the two inclined oculars, a vertical phototube is present at the top of the microscope. The phototube permits attachment of a camera (with or without exposure meter) or a projection device, without interfering with the image seen through the two oculars.

Most student microscopes (with proper adapters) permit attachment of a single lens 35 mm reflex camera to the single ocular that is present. Focusing and determination of exposure can be made through the viewfinder of the camera. Although this is not an ideal arrangement, it often yields adequate photomicrographs.

If possible, a flat field (planar) objective should be used to provide sharp focus in all parts of the field of view. Additional control of focus is provided by focusing the microscope with the aperture diaphragm open. Before taking the photo, the aperture diaphragm should be partially closed to increase the depth-of-field of the lens and provide a better chance of a good focus.

When taking black-and-white photographs, it is best to use panchromatic film because it records all wavelengths. If particular colors are to be emphasized, an appropriately chosen filter may be used.

Color photographs can be taken with a film that is suitable for a tungsten filament light source. It is important that the color temperature of the illuminator match that of the film; this is achieved by proper adjustment of a variable intensity illuminator. In addition it may be necessary to insert a color filter into the microscope system to compensate for differential absorption of light within the microscope system. The particular filter required may be furnished with the microscope or may be determined by viewing a previously exposed and developed color transparency through a variety of filters.

ADDITIONAL READINGS

Hallimond, A. F. 1970. *The Polarizing Microscope*, 3d ed. York, U.K.: Vickers Instruments, 1–49.

Hartshorne, N. H., and A. Stuart. 1969. *Practical Optical Crystallography*, 2d ed. New York: American Elsevier, 109–153.

Jones, N. W., and F. D. Bloss. 1980. *Laboratory Manual for Optical Mineralogy*. Minneapolis: Burgess, 1-1 through 1-9.

Schaeffer, H. F. 1953. *Microscopy for Chemists*. New York: Dover, 1-97.

Wahlstrom, E. E. 1979. *Optical Crystallography*, 5th ed. New York: John Wiley & Sons, 83–106.

2

Light and Matter

> And God said, "Let there be light"; and there was light. And God saw that the light was good; and God separated the light from the darkness.
>
> Genesis 1:3–4

In order to operate the petrographic microscope properly, it is necessary to know something about the nature and behavior of light. Light is radiant energy consisting of electromagnetic waves; that is, both electrical and magnetic components are present. Alternatively, light can be described in terms of light quanta (photons), which behave as particles of matter. It is convenient for our purposes to consider light as transverse waves transmitted through matter by distortions of electronic orbits. We will be concerned only with the electrical aspects of light.

WAVELENGTH AND COLOR

One of the characteristics of wave motion is wavelength. Wavelength is designated by the Greek letter lambda (λ), and is defined as the shortest distance between corresponding parts of a wave, as from one crest to the next or from one trough to the next (Fig. 2-1). Commonly used units of measurement are nanometers (nm) or the obsolete millimicrons (mμ); see Table 2-1 for equivalents. Visible light, that which stimulates our optic nerves, consists of a limited range of wavelengths within the total electromagnetic spectrum that extends from 10^{-7} nm for cosmic rays to 10^{16} nm for long electrical waves (Fig. 2-2). Compared to photographic film or electrical sensing devices, the human eye is sensitive to only a small portion of the electromagnetic spectrum, the wavelengths from about 400 nm to about 770 nm.

Table 2-1.
Equivalency of commonly used units of wavelength

1 nanometer (nm) =	10^{-9} meter (m)
	10^{-7} centimeter (cm)
	10^{-6} millimeter (mm)
	10^{-3} micron (μ)
	1 millimicron (mμ)
	3.937×10^{-8} inch
	10 angstroms (Å)

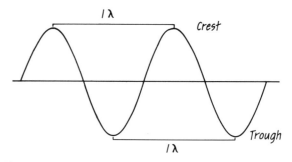

Figure 2-1
A sinusoidal wave produced by combining simple harmonic oscillation with uniform velocity of travel. A wavelength (designated as λ) is defined as the shortest distance between corresponding parts of a wave.

The eye interprets various wavelengths (or frequencies, or energies) of light as different colors (Fig. 2-2). The longer wavelengths near 770 nm

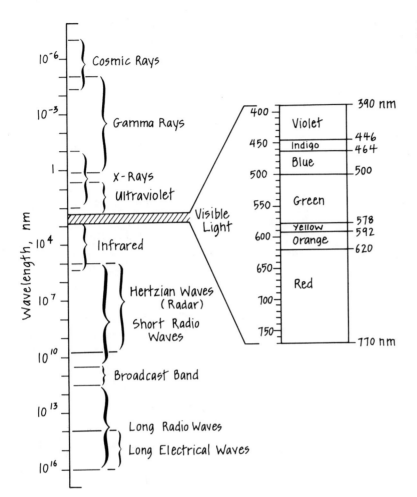

Figure 2-2
The electromagnetic spectrum. The colors associated with visible light are shown in an expanded scale to the right.

are seen as red, and progressively shorter wavelengths are viewed as orange, yellow, green, blue, indigo, and violet. Sunlight and light from incandescent (tungsten filament) bulbs represent a combination of all of these wavelengths, but not in equal proportions; the combined color is seen as white light. Sunlight has a greater proportion of the shorter wavelengths and has a bluish hue in the northern sky, whereas an incandescent bulb has a preponderance of the longer wavelengths and has a yellow-orange hue (Fig. 2-3).

Light produced by waves of a single wavelength is called *monochromatic*. *Polychromatic* light contains a variety of wavelengths. If two colors of light combine to produce white light, they are

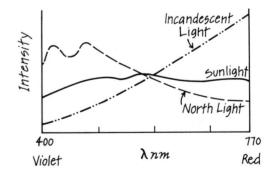

Figure 2-3
The wavelength distribution of sunlight, light from a north-facing window, and an incandescent light.

8 Chapter 2: Light and Matter

called complementary colors. Several complementary combinations exist, such as green and violet, yellow and blue-violet, or red and blue-green. A single combination of the three primary colors (red, green, and blue-violet) also produces white light. Alternatively, light of two different wavelengths (such as blue and yellow) may be superimposed on the same surface to produce a third color (green). The particular color seen by the eye might be produced by light of a single wavelength or by a mixture of light of several wavelengths; the human eye cannot distinguish between the two situations.

FREQUENCY, ENERGY, AND VELOCITY

Electromagnetic waves are emitted from their source at a certain number of vibrations (wavelengths) per unit time; this rate of wave emission is called *frequency*. The frequency of a light wave does not change when the light passes from one material to another, as this value is determined by the emitting source. Consider the following relation:

$$E = h\nu \tag{1}$$

where E is energy, h is Planck's constant, and ν is frequency. It follows from this equation that the frequency of a light wave is directly proportional to its energy. The frequency (and hence the energy) of a light wave determines the response of the material being illuminated.[1]

Another standard relation is

$$V = \nu\lambda \tag{2}$$

where V is velocity, ν is frequency, and λ is wavelength. Let us consider how this relation can be used.

---

1. This follows from the relationship between equations (1) and (2), $E = h\nu$ and $V = \nu\lambda$. As $\nu = V/\lambda$, therefore $E = h\nu$ can be transformed to $E = h(V/\lambda)$, so that $E\lambda = hV$. As hV is constant for any chosen medium, E and λ must have an inverse relationship.

The velocity of light in free space is 2.997925×10^{10} cm/s (186,282 miles per second). This measured velocity is independent of wavelength; that is, all colors of light move at the same speed. It follows from equation (2) that in free space (vacuum) the frequency of emission must be inversely related to wavelength. It is, in fact, the frequency of the light rather than the wavelength that causes our eyes to see a particular color.

It is well known that the velocity of light in media other than free space is less than 2.997925×10^{10} cm/s. For example, light moves through water at only 2.25×10^{10} cm/s (133,000 mi/s). As the frequency of the light is controlled by the source and cannot change, it follows from equation (2) that the wavelength must decrease in a medium other than free space, in direct relation to the velocity decrease. As the frequency is unchanged, the color of the light is unchanged within the medium. Upon returning to free space, the velocity of light rises again and the wavelengths resume their original values.

PLANE-POLARIZED LIGHT

A ray of light is considered to be nonpolarized or "normal" if it vibrates equally in all directions perpendicular to its direction of travel. If the light is restricted in its vibrational direction, it is said to be *polarized*. One type of restriction results in *plane-polarized light*. Here, the light is constrained to vibrate in a single plane perpendicular to its direction of travel. Such a situation is shown in Figure 2-4: the plane-polarized light, traveling from left to right, is restricted to a plane of vibration parallel to the page. This plane is the *vibration plane*; the direction of displacement vectors (shown as vertical arrows) within the plane is called the *vibration direction*. Note carefully that the vibration direction does not correspond to the *propagation direction* (A to B) of the light ray.

Plane-polarized light can be created by reflection of nonpolarized light from a nonmetallic surface (such as glass or an insulating dielectric material). This was discovered by Etienne Malus in

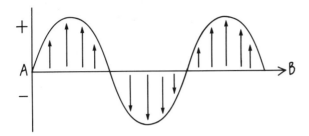

Figure 2-4
Wave motion confined to a single plane (the plane of the page). The plane in which the wave motion occurs is the vibration plane, and the direction of wave displacement (arrows within the vibration plane) is the vibration direction. The direction of propagation is from A to B.

1808. Light at inclined incidence to a reflecting surface is often both reflected and refracted (that is, the refracted beam changes propagation direction when crossing the interface) (Fig. 2-5). David Brewster in 1811 discovered that when the angle between the reflected and refracted rays is 90°, the reflected rays are completely plane-polarized and the refracted rays are partially plane-polarized. The vibration direction of the reflected rays is parallel to the surface of reflection (shown in Figure 2-5 as small circles); the refracted rays are partially plane-polarized in the plane containing the incident and refracted rays (with vibration direction shown as short lines in the plane of the paper).

The explanation for this special case is based on the fact that the propagation direction of the reflected rays is parallel to the vibration direction of the refracted ray. Electromagnetic theory requires that no energy can be radiated in a wave's vibration direction,[2] hence the vibration direction of the refracted beam (within the plane of the section) eliminates those vibration directions of the reflected beam that are also within the plane of the section.

Most artificial plane-polarized light is generated not by reflection but by absorption. Some crystalline compounds transmit light rays vibrating in one direction while absorbing those rays

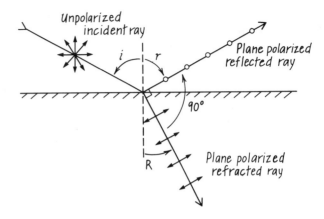

Figure 2-5
A beam of nonpolarized light encountering an interface at an oblique angle may undergo both reflection and refraction. If the angle between the reflected and refracted beams is 90°, the reflected beam is completely plane-polarized; the vibration direction (shown by small circles) is parallel to the reflecting surface, into and out of the page. The refracted beam is partially plane-polarized in the plane containing the incident and refracted rays (shown by short inclined lines).

vibrating in all others. This property, called *differential absorption*, is well developed in tourmaline, which transmits only light rays vibrating in the c crystallographic axis direction. If two tourmaline crystals are arranged so that their c axes are perpendicular to each other (and to the light's direction of propagation), virtually no light is transmitted (Fig. 2-6).

The polarizing devices in most modern petrographic microscopes use Polaroid, a trade name for a plastic film produced by the Polaroid Corporation of Cambridge, Massachusetts. Polaroid consists of long-chain organic molecules (usually polyvinyl alcohol) that are aligned by stretching and treated with iodine. It is a crystalline compound with strong differential absorption that converts transmitted nonpolarized light into plane-polarized light.

---

2. This restriction does not apply to some crystalline materials, as discussed in Chapter 4.

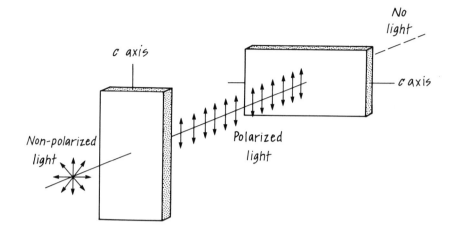

Figure 2-6
A plate of tourmaline absorbs all light vibrations except those vibrating in a plane that includes the *c* axis. Nonpolarized light entering such a crystal emerges as plane-polarized light. If this light encounters a second tourmaline crystal whose *c* axis is at 90° to the first, the light is almost completely absorbed.

LIGHT TRANSMISSION THROUGH ISOTROPIC AND ANISOTROPIC MATERIALS

Materials that have identical optical properties in all directions are *optically isotropic:* the velocity of light is constant in all directions of transmission. Gases, liquids, and noncrystalline solids such as glass or the essentially amorphous mineraloids fall into this category; isometric crystalline materials (such as the minerals fluorite and halite) are also isotropic, because they have equivalent atomic packing in the three perpendicular crystallographic axial directions.

Materials whose optical properties are different in different directions are said to be *optically anisotropic;* all hexagonal, tetragonal, orthorhombic, monoclinic, and triclinic materials are in this category. Within these materials, the velocity of light varies as a function of propagation direction because of directional differences in atomic arrangement.

Color

The electrical component of the electromagnetic light wave mainly affects the electrically charged electrons of the atoms that compose solids. The frequency of light waves (5×10^{14} vibrations per second) is so high that there is little effect on the electrically charged atomic nuclei; almost all of the mass of an atom is concentrated within the nucleus, and its inertia is relatively high. The electrons, however, are light and mobile, and hence easily displaced from their orbital equilibrium positions. Such displacements affect the color of a mineral as well as the velocity and type of polarization of a transmitted light beam.

The color of most minerals results from the depletion of certain wavelengths from the full spectrum. A mineral such as olivine, $(Mg,Fe)_2SiO_4$, is considered to be transparent, in that it transmits visible light. However, not all of the visible light is transmitted. Consider a beam of sunlight traveling through a crystal of Mg-rich gem-quality olivine (peridot), prized for its green color. As the white sunlight penetrates the peridot, wavelengths near the long and short wavelength ends of the spectrum are absorbed, with the result that an incomplete (depleted) spectrum passes through the mineral. This depleted spectrum combines to yield the green color of the mineral.

Other minerals do not transmit any light and are classified as opaque. Sunlight encountering such materials is partially absorbed and turned into heat; the balance of the light is reflected from the surface. The reflected light consists of a depleted spectrum, which again combines to yield the color of the mineral. Thus the color of a substance may combine reflection, absorption, and transmission of the various wavelengths of light that impinge upon it.

The absorption of light by a crystal often occurs by displacement of electrons. Electrons may respond to the energy of a light wave by being displaced or excited to higher than normal energy levels. Each of these electron-jumps (transitions) from the lowest (ground) energy state to a specific higher level is related to a particular amount of energy; in other words, light of a particular frequency is absorbed. After a short time the excited electrons return to their ground state, usually by a series of transitions to lower energy levels, each transition emitting a quantum of light of specific energy. If the emitted light has an energy in the visible region, it is called *fluorescence*. The initial absorbence, however, gives rise to the color observed in most minerals. This is the case for peridot, for instance; the energy it re-emits is in the nonvisible infrared wavelengths.

Alternatively, consider the red color of ruby. The chromium atom in ruby (approximately $Al_{1.9}Cr_{0.1}O_3$) absorbs light corresponding to violet and yellow-green (by excitation of the outermost d-shell electrons to higher energy states); the depleted spectrum that is transmitted through the ruby is strong in the unabsorbed red wavelengths. The red color is enhanced by red fluorescence as the absorbed light is re-emitted by the excited electrons.[3]

Atomic Polarization by Light

If the energy state of the incident light and that required for an electronic transition do not correspond, the electrons are not excited to higher energy levels; instead, the orbits of the most weakly held electrons are distorted. Consider what happens to a single atom kept between two surfaces having electrically opposite charges, as in a capacitor (Fig. 2-7). The positively charged nucleus

---

3. A fuller discussion of the causes of color is given by Nassau (1980).

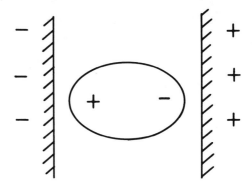

Figure 2-7
Atomic polarization (elliptical distortion) of an atom between electrically charged surfaces. The positive nucleus is attracted toward the negative surface, and the negative electrons are attracted toward the positive surface.

is somewhat attracted to the negatively charged surface, while the negatively charged electronic distributions are somewhat elliptically distorted, shifting their center of gravity closer to the positive plate. The atom is then said to be polarized and behaves like an electric *dipole*. Note carefully that "polarization" as used here has (unfortunately) a different meaning from its application to light.

The extent of polarization of the atom—the *dipole moment*—depends upon both the intensity of the superimposed electric field and the character of the atom itself. Large atoms, which have loosely attached electrons in their outermost orbits, are generally more easily polarized than smaller atoms (whose electrons are more tightly held). Another influencing factor is the valence state. A positive ion (cation) contains fewer electrons than a neutral atom of the same element, and these are held more tightly by the nucleus than the outermost electrons in a neutral atom; consequently cations are only weakly polarizable. A negative ion (anion) contains more electrons than does a neutral atom and thus is more easily polarized.

In most crystalline materials, each atom is surrounded closely by a variety of neighboring

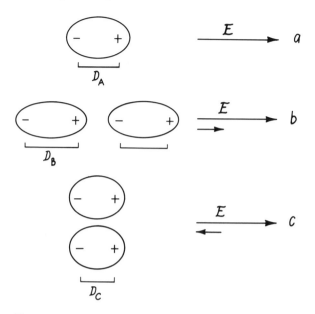

Figure 2-8
(A) A single atom is polarized because of the electric vector **E**. The extent of polarization is indicated by length D_A. (B) Two adjacent atoms aligned parallel to the electric vector are enhanced in their polarization, thus $D_B > D_A$. (C) Two atoms aligned normal to the electric vector have reduced polarization, thus $D_C < D_A$.

atoms of diverse size and charge. This results in a directionally nonuniform environment. Certain regions around most atoms contain slightly more (valence) electrons than others, due to the formation of chemical bonds. The bonding electrons are more polarizable than the inner shell electrons; that is why in many crystalline materials, certain directions are more easily polarized than others.

As a very simple model of the effects of light on atoms within a crystal, consider Figure 2-8. In part A, a single atom is polarized by the oscillating electric vector **E** of a light wave. The extent of polarization of the atom is indicated by the length D_A, which is a measure of the distance between the center of the nucleus and the center of gravity of the electrons. At a later instant in time, the oscillating electric vector causes the atom to be polarized in an opposite (180°) direction; thus the light and the atom oscillate synchronously. Part B shows two similar adjacent atoms, oriented parallel to the electric vector. Each atom is polarized by the electric vector to the same extent as in part A, but this polarization is enhanced by a neighboring atom (as opposing charges in adjacent atoms are aligned by the polarization); thus the polarizability is increased over part A, as shown by the extra vector, and D_B is greater than D_A. A longer chain of atoms would be still more polarizable.

In part C, two adjacent atoms are arranged with their centers normal to the electric vector. Each atom tends to be polarized to the same extent as in part A, but this polarization is reduced by the presence of an adjacent atom (as the like charges repel each other); thus polarizability is decreased as compared to part A, by the length of the small vector, and D_C is less than D_A. From this example we can see why polarization is enhanced when the electric vector is parallel to directions in the crystal that contain larger atomic populations.

Let us transfer this understanding to a more realistic crystal. A schematic cross section of a crystal structure of a phyllosilicate mineral is shown in Figure 2-9. Densely packed tetrahedral and octahedral sheets are shown in gray, and circles represent cations that hold the sheets together. The valence electrons that bind the atoms together are concentrated within the sheets. Let a beam of nonpolarized light impinge on the crystal straight down onto the page. The vibration planes (of the electrical portion) of the ray, a few of which are shown schematically below as double-headed arrows, are perpendicular to the ray path. Light vibrating in the plane AA is parallel to the plane in the crystal that contains the greatest number of atoms. Due to the proximity of atoms and easily polarized valence electrons within this plane, the dipole moment in this direction reaches a maximum value. Vibrational plane BB, at right angles to AA, is parallel to a direction of minimum polarization.

In almost every randomly chosen plane through

an anisotropic crystal, there exist mutually perpendicular directions of maximum and minimum polarization. If the vibration direction of the incident light is not either of these directions, it is resolved into these two vibrational components. Hence vibration CC in Figure 2-9 is resolved into components that vibrate in the planes AA and BB. Light vibrating in such mutually perpendicular planes is called here *doubly polarized light*.

Isotropic substances do not make transmitted light doubly polarized. Instead, the light's original state of polarization is retained because isotropic materials do not possess differences in atomic packing density in two mutually perpendicular planes. In isometric crystals, for example, the atomic arrangement is identical along the a_1, a_2, and a_3 crystallographic reference axes.

INTERFERENCE

Two different light waves can be combined to produce a single composite vibration. Consider first the combination of two plane-polarized monochromatic light waves within the same vibration plane. The resultant single vibration will have characteristics that depend upon how the two original waves are displaced relative to each other. The general rule is that the displacement of the resultant wave at any point is equal to the vector sum of the displacements of the two component waves. For example, if two superimposed waves of the same wavelength are in phase, then all parts of the two waves correspond in displacement, and the crests and troughs of one match the crests and troughs of the other (Fig. 2-10A). The superposition of the two waves is additive at all points, and the resultant wave has higher crests and lower troughs than either of the two original waves. When superimposed waves are exactly $\frac{1}{2}\lambda$ out of phase, the crests of one wave correspond to the troughs of the other, the vector sum of the two displacements at any point is zero, and no resultant wave is produced (Fig. 2-10B). If the two waves are out of phase by an amount other

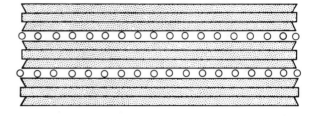

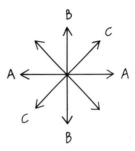

Figure 2-9
Schematic cross-section of a phyllosilicate crystal structure. Close-packed layers of atoms are represented as gray bars. Interlayer atoms (circles) link the close-packed layers together. Electric vectors of a light ray, traveling perpendicular to the page, are shown below as double-headed arrows. Vector direction AA corresponds to the direction of maximum atomic polarization within the structure. Vector direction BB, at right angles to AA, corresponds to the direction of minimum atomic polarization.

than $\frac{1}{2}\lambda$ or 1λ, the resultant wave has the same wavelength as the two original waves, but the position of crests and troughs does not correspond to either of the two original waves (Fig. 2-10C). Superposition of two waves of different wavelengths usually produces a resultant wave of irregular wavelength and displacement (Fig. 2-10D).

When the amplitude (displacement) of a wave is increased or decreased, it is said to be *amplified* or *attenuated*, respectively. Complete elimination of a wave is *nullification*. The *intensity* (brightness) of a light wave is proportional to the square of the amplitude.

Consider superposition of light waves within solids and liquids. The electric component of inci-

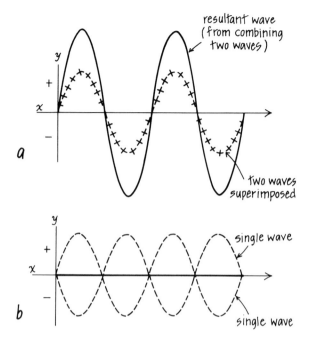
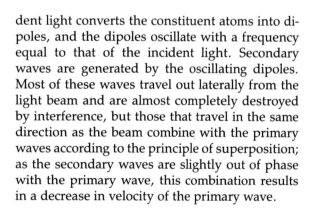
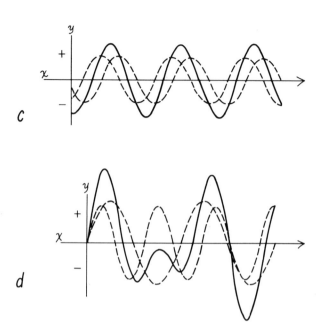

Figure 2-10
The superposition of two plane-polarized light waves on the same vibration plane. (A) The two waves correspond in both wavelength and displacement. When superimposed, the resultant wave has the same wavelength as the two originals, but the displacement is doubled. (B) The two waves correspond in wavelength but have opposite displacements. Vector addition of the displacements results in elimination of the resultant wave. (C) The two waves have the same wavelength, but their displacements neither correspond nor are opposite. The resultant wave has the same wavelength, but its displacement positions are different from the two original waves. (D) The two waves differ in wavelength, displacement positions, and amplitude. Superposition results in a wave of variable wavelength and nonregular displacement.

dent light converts the constituent atoms into dipoles, and the dipoles oscillate with a frequency equal to that of the incident light. Secondary waves are generated by the oscillating dipoles. Most of these waves travel out laterally from the light beam and are almost completely destroyed by interference, but those that travel in the same direction as the beam combine with the primary waves according to the principle of superposition; as the secondary waves are slightly out of phase with the primary wave, this combination results in a decrease in velocity of the primary wave.

Within anisotropic materials, the decrease in velocity of light is related to the directions of maximum and minimum polarization of the crystal. The greatest decrease in velocity is parallel to the plane of maximum polarization, and the least is in the plane of minimum polarization. As $V = \nu\lambda$, it follows that the two vibrations not only have two different velocities, but two different wavelengths as well.

Another superposition of light waves occurs upon emergence from an anisotropic crystal, when the waves from the two perpendicular vibration planes combine to form a single resultant wave. This occurs because the crystallographic constraints of the crystal are no longer present. Depending upon the positions of crests and troughs of the two waves, the resultant wave may be plane-polarized, circularly polarized, or elliptically polarized. In circularly and elliptically polarized light, the vibration direction rotates about the propagation direction (Fig. 2-11). In *circularly polarized light*, the amplitude of the waves is constant in all directions; in *elliptically polarized light*, the wave amplitude changes with direction in an elliptical pattern. This subject is covered in more detail in Chapter 5.

ADDITIONAL READINGS

Bouma, B. J. 1947. *Physical Aspects of Colour*. Eindhoven, Neth.: N. V. Philips, 312 pp.

Ditchburn, R. W. 1952. *Light*. London: Blackie and Son, 680 pp.

Ghatak, A. K. 1972. *An Introduction to Modern Optics*. New York: McGraw-Hill, 368 pp.

Longhurst, R. S. 1957. *Geometrical and Physical Optics*. London: Longmans, Green, 534 pp.

Mueller, C. G., M. Rudolf, and the Editors of *Life*. 1966. *Light and Vision*. New York: Time, Inc., 200 pp.

Shurcliff, W. A., and S. S. Ballard. 1964. *Polarized Light*. Princeton: D. Van Nostrand, 144 pp.

Zoltai, T., and J. H. Stout. 1984. *Mineralogy: Concepts and Principles*. Minneapolis: Burgess, 263–291.

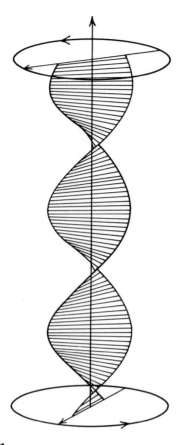

Figure 2-11
A schematic representation of an upward-traveling circularly polarized ray of light. The plane of polarization (vibration plane) rotates in a counterclockwise direction, forming a corkscrew-like pattern. The changing vibration directions are shown as thin horizontal lines within the plane of polarization. The amplitude remains constant for circularly polarized light; for elliptically polarized light, the amplitude changes with direction in an elliptical pattern.

3

Examination of Isotropic Substances

Anyone can steer the ship when the sea is calm.
PUBLILIUS SYRUS

An isotropic substance transmits monochromatic light at a constant velocity that is independent of direction. If the substance is subjected to different values of pressure and temperature, or different wavelengths of incident light, the velocity of monochromatic light within it changes, but remains independent of direction. The velocity of light within a substance is easily determined with a petrographic microscope by measuring the index of refraction (commonly symbolized as n).

INDEX OF REFRACTION

The index of refraction of a substance is a number derived by dividing the velocity of light (V) within the substance into the velocity of light in vacuum:

$$n = V_{vacuum}/V_{substance}$$

As the velocity of light in all substances is less than that of light in vacuum, refractive indices are numbers greater than 1. The velocity of light in air is almost the same as the velocity in vacuum, and is considered equivalent in most situations, as n_{air} is 1.0003. A unique index of refraction is often defined at a fixed wavelength (such as sodium light of $\lambda = 589$ nm) under standard pressure-temperature conditions.

The decrease in velocity of light that occurs when a beam of light passes from air into a solid or liquid is accompanied by a decrease in wavelength. From the relation $V = \nu\lambda$, it follows that each wavelength of light travels through solids and liquids at a different velocity. As the frequency (ν) of each wavelength is fixed by the emitting source, the light of long wavelength has greater velocity than the light of short wavelength within a solid or liquid. As the index of refraction is a ratio of velocities of light, it follows that n varies as a function of wavelength. This variation is different for different substances; thus, in isotropic substances, the indices of refraction of longer wavelengths are lower than those of shorter wavelengths. This variation of n as a function of wavelength is called *index dispersion* (Fig. 3-1).[1]

The amount of index dispersion is a measurable parameter. Liquids generally show higher dispersion than solids. When precise characterization of index dispersion is required, it is usually given as the *coefficient of dispersion:* the difference in indices of refraction measured at wavelengths of 486.1 nm and 656.3 nm.[2]

---

1. Optical dispersion refers to any kind of variation of an optical property with frequency (or wavelength). Other types of dispersion are discussed in Chapter 9.
2. These wavelengths correspond to the easily observed F and C Fraunhofer lines.

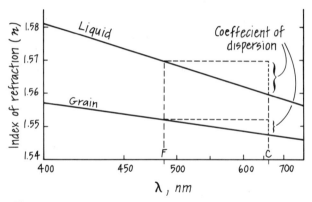

Figure 3-1
Index dispersion curves for typical liquids and solids. The abscissa scale, based on the Hartmann equations (see Wahlstrom, 1979, pp. 113–114), is used in order to yield straight, rather than curved, dispersion curves. Index dispersion curves of liquids generally have greater negative slopes than those of solids.

The measurement of n is usually based upon the relation between the angles of incidence and refraction of light at an interface between two materials. Assume two isotropic media in coherent contact and a monochromatic light source. In Figure 3-2 the incident ray of light travels from the upper left in medium 1 and encounters medium 2 obliquely. If the velocity of light in the second material is identical to that in the first, the ray crosses the interface without deviation. If the velocity of light is different in the two materials, the incident ray may be reflected or refracted (or both) at the interface. A reflected ray bounces off the surface and remains in medium 1; its angle of incidence (angle i in Fig. 3-2) is the same as its angle of reflection (angle r).

Alternatively, the incident ray may cross the interface and enter the second material. If the ray then deviates from its original direction of travel, it is called the refracted ray. The angle of refraction R (Fig. 3-2) and the angle of incidence i are not equal: if R is less than i, the velocity of light in the second material is less than in the first, and if R exceeds i, light has a greater velocity in the second material.

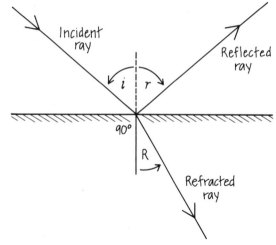

Figure 3-2
A light ray is both reflected and refracted at an interface between two nonopaque media. The incident (oncoming) ray in the upper medium encounters the interface with an angle of incidence i. The reflected portion of the ray has an angle of reflection, r, that is always equal to the angle of incidence, i. Light rays that cross the interface into the lower medium are refracted because of a change in velocity between the two media. The angle of refraction is angle R.

The relationship between the incident and refractive angles and the velocity of light was explained by Willebrord Snell in 1621. Figure 3-3 shows two parallel light rays refracted across an interface between medium 1 (with index of refraction n_1) and medium 2 (with index of refraction n_2). At some instant in time, rays 1 and 2 have reached A and B respectively. The front of the light waves (wave front) is shown as the line AB. At a later time, ray 1 reaches C in medium 2 just as ray 2 arrives at the interface at D. The wave front in medium 2 at this later time is CD. The distance BD can be taken as a measure of the velocity (V_1) in medium 1, and the distance AC can be taken as a measure of the velocity (V_2) in medium 2.

Wave fronts AB and CD are normal to the ray paths in the two media. As triangles ABD and ACD are therefore right triangles, it follows that

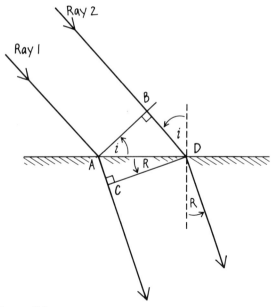

Figure 3-3
Two parallel light rays are refracted across an interface between two nonopaque media. The line AB represents a wave front within the upper medium. A wave front CD within the lower medium is shown after refraction across the interface. The wave fronts are normal to the ray paths in each medium. The lengths BD and AC are proportional to the velocities of light in the two media. The angle of incidence i, shown at point D, is equivalent to angle BAD. The angle of refraction R, shown at D, is equivalent to angle ADC.

angle BAD is equal to the angle of incidence i, and angle ADC is equal to the angle of refraction R. Since

$$\sin i = \frac{BD}{AD} \quad \text{and} \quad \sin R = \frac{AC}{AD},$$

then

$$\frac{\sin i}{\sin R} = \frac{BD}{AC}.$$

Recalling that

$$BD = V_1 \quad \text{and} \quad AC = V_2,$$

then

$$\frac{\sin i}{\sin R} = \frac{V_1}{V_2}.$$

As

$$n_1 = \frac{V_{vacuum}}{V_1} \quad \text{and} \quad n_2 = \frac{V_{vacuum}}{V_2},$$

substitution results in the following general version of Snell's law:

$$\frac{\sin i}{\sin R} = \frac{n_2}{n_1}$$

If the material 1 is a vacuum or air, such that n_1 is 1, then

$$\frac{\sin i}{\sin R} = n_2$$

As Snell's law contains four variables, knowledge of any three permits calculation of the fourth. For example, if a beam of light in air ($n = 1.00$) enters water ($n = 1.33$) at an incident angle of 45°, it is easily calculated that the angle of refraction is 32°. This equation also reveals that light having perpendicular incidence to a surface ($i = 0°$) will cross the surface with angle $R = 0°$.

In addition, the equation permits calculation of the *critical angle:* the incident angle at which the refracted beam (traveling from a higher to a lower index medium) has an angle of refraction of 90°. Here the refracted beam travels along the interface between the two media.

DETERMINATION OF INDEX OF REFRACTION

The index of refraction of an isotropic solid may be determined by observing the effects of refraction. When working with crushed grains, a set of standardized immersion liquids (whose indices of refraction are precisely known) is used. A set of liquids ranging in value from 1.43 through 1.74 (in 0.01 intervals) is adequate for most purposes. For more detailed studies, sets having more precise intervals—such as 0.005 or 0.002—can be used. Sets of immersion liquids can be purchased from many scientific supply houses and are available in most microscope laboratories. Alternatively, sets of liquids may be made with binary mixtures of reagents. Frequently used liquids are ethyl butyrate ($n = 1.381$), mineral oil ($n = 1.470$), α-monochloronaphthalene ($n = 1.633$), and diiodomethane ($n = 1.739$).[3] Care should be taken to verify that the immersion liquids have been calibrated within the last few months before using them. This is usually done with an Abbe refractometer (with an accuracy of ± 0.0002).

The sample is prepared for examination by pulverizing and sieving (in order to achieve a consistent fragment size). Grains trapped between 100- and 200-mesh sieves are best for most purposes. A few dozen grains are placed on a glass slide, then covered with a cover glass (usually about 0.17 mm thick). The grains are then immersed in a few drops of an immersion liquid (usually between 1.55 and 1.65), an operation done by using the dropper rod from the immersion liquid bottle (Fig. 3-4). Touching the dropper rod to the end of the cover glass allows the liquid to be drawn by capillarity between the cover glass and the slide. An alternative method is to place the liquid on the slide first. The grains are then sprinkled in the liquid, and the cover slip is placed on top; this technique avoids any possibility of contaminating the liquid with mineral grains or coating the top of the cover glass with liquid, but has the disadvantage of obscuring the number and position of

---

3. Care should be taken to choose immersion liquids that do not dissolve or react with the material under observation. The vapor of certain immersion liquids has been reported to be somewhat toxic; extended usage of immersion liquids should be carried out in a well-ventilated area. Data on the indices of refraction of liquids and their toxicity are given in the *Handbook of Chemistry and Physics*.

Figure 3-4
Preparation of a sample mount. Grains are placed on a glass slide and covered with a cover glass. An immersion liquid is applied to the edge of the cover glass with a dropping rod. The liquid is drawn under the cover glass by capillarity. It may be necessary to gently push or slide the cover glass in order to disperse the grains within the liquid.

the grains. If most of the grains are not immersed in the liquid or are bunched together, the cover glass should be gently moved or pressed using a clean tool, not the fingers. Fingerprints on the cover glass or slide tend to obscure the image.

The slide is then put on the microscope stage and brought into focus with the low-power objective. This is best done by first bringing the objective close to the slide while looking at the sample from the side (not down the tube), then bringing it into focus by raising the objective or depressing the stage. This technique avoids crushing the slide with the lens. For the same reason, do not use the higher power objectives for initial focusing: focus the grains at low power, then bring the higher power objectives into use as necessary. The strategy is to make a series of such grain mounts, in order to match the grains with a liquid of identical or very close index of refraction.

Relief

Typically, in the first grain mount, there is an appreciable difference between the indices of the

Figure 3-5
(A) A fragment showing high relief, resulting from a large difference in index of refraction between the grain and the immersion liquid. (B) A fragment showing low relief; the difference in indices of refraction between the grain and the immersion liquid is small.

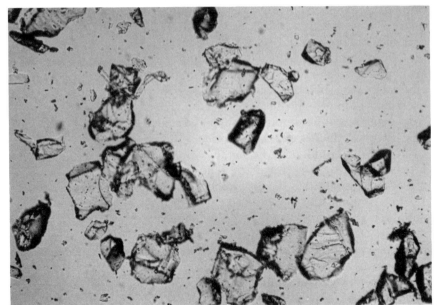
a

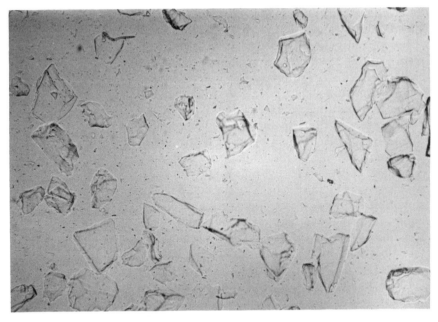
b

grains and immersion liquid. If this is the case, the vertical rays of light coming from the substage into the sample are strongly refracted as they enter and leave the fragments. This produces an unequal distribution of light within and above the fragments (Fig. 3-5A). The unequal distribution of light gives the impression that the grains have a relatively high relief—that is, they are easily observed, and have sharp shadows on their borders.

On the other hand, if the grains and liquid have about the same index, refraction effects are small, and the grains have low relief (Fig. 3-5B); they appear flat and featureless. An exact match between colorless grains and liquid (using a monochromatic light source) results in the grains being invisible (except for inclusions and impurities); even colored grains appear as only faint patches of color with indistinct boundaries.

The Becke Line

Assume that the first grain mount shows moderate to high relief. The next task is to determine whether the grains are higher or lower in index than the immersion liquid. This is done using a test devised by the German mineralogist F. Becke in 1893.

If the grains have a higher index of refraction than the immersion liquid, light rays entering the grains are refracted so as to produce a slight convergence above the grain (Fig. 3-6). This effect is most pronounced at the grain edges (due to the high angle of incidence). As determined by Becke, a band of concentrated light can be observed if the light source is darkened by partially closing the aperture diaphragm and if the microscope is focused slightly above the grain. A very thin grain edge is best for this observation. By raising the microscope tube or lowering the stage, the grain goes somewhat out of focus, and a weak band of light can be seen moving from the grain edge into its interior. This area of concentrated light, which follows the outline of the grain edge, is called a *Becke line*. It moves into the higher index material as the distance from grain to lens is increased, and in this case the *Becke test* shows

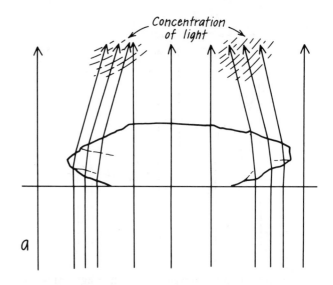

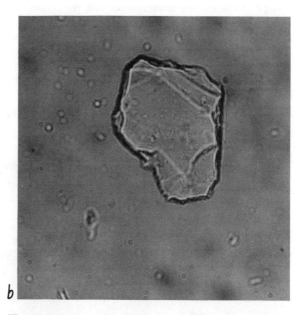

Figure 3-6
(A) When the index of refraction of the fragment is higher than that of the immersion liquid, refraction tends to concentrate the light rays above and toward the center of a lenticular grain, particularly at the grain's inclined edges. (B) Areas of light concentration above the grain, the Becke line, move inward from the grain edge as the microscope tube is raised or the stage is lowered. The aperture diaphragm must be partially closed for this observation.

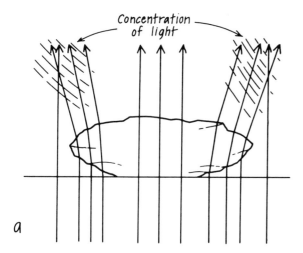

Figure 3-7
(A) When the index of refraction of the fragment is lower than that of the immersion liquid, refraction tends to concentrate the light rays above and away from the center of the grain, particularly at the edges. (B) Areas of light concentration above the grain, which constitute the Becke line, move outward from the grain edge as the microscope tube is raised or the stage is lowered. The aperture diaphragm must be partially closed for this observation.

that the solid has a higher index of refraction than the liquid. In order to achieve a match between the solid and liquid, a higher index liquid must be chosen.

Assume now that a second mount is made, and the liquid chosen has a higher index than the grains. When the Becke test is made, the Becke line is seen to move away from the grain edge into the liquid (Fig. 3-7). And, of course, if the grain is brought back into focus, the Becke line returns to the grain edge.

After it has been established that the unknown solid lies between the indices of refraction of two different immersion liquids, it is then easy to make successive mounts until the solid and the immersion liquid match. It is here that observation of grain relief is important, as the relief furnishes a clue to the magnitude of difference between the index of the grain and the liquid.

The Double Becke Line

If monochromatic light is used, a colorless grain becomes totally invisible when a perfect match is achieved. Alternatively, if polychromatic light is

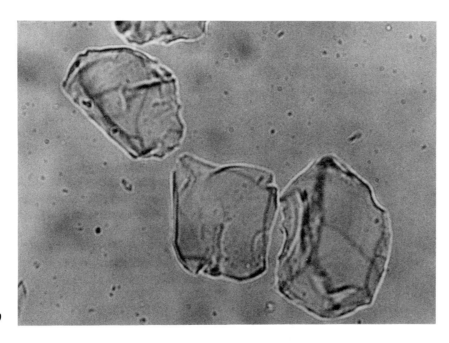

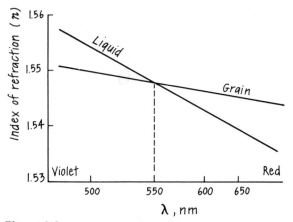

Figure 3-8
Index dispersion curves of a liquid and solid, shown as straight lines for simplicity. A perfect match between the liquid and the solid occurs where the two curves intersect, in the middle of the visible spectrum. For longer wavelengths the solid has a higher index than the liquid, and for shorter wavelengths the liquid has the higher index. An inward-moving red Becke line has the same intensity as an outward-moving violet Becke line.

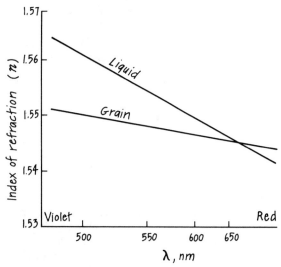

Figure 3-9
When index dispersion curves of a liquid and a solid cross near one end of the visible spectrum, the two Becke lines produced are of unequal intensity. In this example, the violet Becke line is more intense than the red Becke line because it comprises the major portion of the visible spectrum.

used, a double Becke line is seen when a match or near-match between the liquid and grain is achieved. A double Becke line consists of two colored Becke lines, one of which moves into the grain and the other out of the grain when the grain-to-lens distance is varied.

This effect is caused by index dispersion. Figure 3-8 shows typical index dispersion curves for a solid and a liquid (plotted as straight lines on Hartman dispersion paper); the liquid, as is usual, has a greater coefficient of dispersion than the solid. As the curves are not parallel, it is obvious that a liquid-solid index match can only be made at a particular wavelength. In the figure, a match is obtained at $\lambda = 550$ nm, where the two curves intersect. For wavelengths greater than 550 nm the solid has a higher n than the liquid; for wavelengths less than 550 nm the liquid has the higher n. The result is that the Becke test produces two Becke lines rather than one, and the grains appear to be surrounded by dim and diffuse colored rims.

A Becke line that moves into the solid consists of the longer wavelengths and has a red-orange hue; simultaneously a blue-violet Becke line moves into the liquid. If the intensity (brightness) of these two Becke lines is about equal, it can be assumed that the two dispersion curves intersect at or near the center of the visible spectrum. This is the best match that can be obtained with a polychromatic light source. Note also that the dispersion curves of various solids and liquids are differently inclined; consequently different solid-liquid pairs produce double Becke lines to different extents.

If two dispersion curves intersect near one end of the visible spectrum, two Becke lines are produced, but they are of different intensity. In Figure 3-9, the two curves intersect near the long wavelength end of the spectrum. For most of the wavelength range, the liquid has higher index

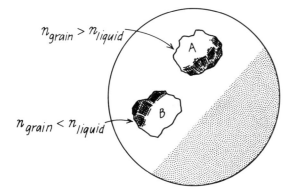

Figure 3-10
The oblique illumination method. The lower right part of the field is shadowed by covering about half of the light from below. Grain A, having a higher index than the surrounding liquid, is shadowed on its "southeast" side. Grain B, having a lower index than the surrounding liquid, is shadowed on its "northwest" side. Note that this relationship is reversed in some microscopes.

than the solid; therefore the blue-violet Becke line, which moves into the liquid, is more intense (as it contains a greater part of the spectrum) than the inward-moving Becke line from the red end of the spectrum. The blue-violet line is also more nearly white.

In such a situation, the movement of the more intense Becke line is a guide to the choice of the next immersion liquid. In this case, as the strong Becke line has moved into the liquid (indicating that the liquid has a higher n over most of the visible spectrum), the next immersion liquid chosen should be slightly lower in index.

An alternative, but related, approach is to notice the colors that enter the grain when a close match is achieved. If the grain has a slightly higher index than the liquid, a lemon-yellow line enters the grain during the Becke test; if the grain has the lower index, a reddish-orange or brown color enters the grain during the Becke test. By means of this trial-and-error approach, it is possible to determine the index of refraction of an isotropic material in a few minutes.

The Oblique Illumination Method

In addition to the use of Becke lines, a second, less accurate, technique can be used for index determination: the *oblique illumination method*. A piece of paper or cardboard is placed so as to cover about half of the light emerging from the base of the microscope. This is seen as a half shadow in the field of view (with use of a medium-power objective and open aperture diaphragm) (Fig. 3-10). Most of the grains appear to be half-shadowed. In general, if the shadowed sides of the grains appear on the same side as the shadowed part of the field, the grains have a higher index than the surrounding liquid. If the shadowed sides of the grains appear on the side opposite the shadowed part of the field, the grains have a lower index than the surrounding liquid. But a caution: some microscopes have lens systems that reverse this reaction, so *always* be sure to determine the correct pattern for your microscope before using this technique.

Determining n in Thin Sections

A thin section is a slice of rock that has been sawed and ground to a thickness of 0.03 mm; the rock slice is glued, commonly with epoxy ($n = 1.540$–1.555), Canada balsam ($n = 1.534$), or Lakeside 70 ($n = 1.54$), to a glass slide. A thin cover glass is commonly glued on the top of the rock slice; if this is done, the precise index of refraction of a mineral cannot normally be determined.[4] Because minerals within a rock slice are surrounded by other minerals or the mounting medium, immersion liquids cannot be used for precise index determination; however, the Becke test works between two solids as well as a solid and a liquid. Consequently, it is possible to determine comparative indices between any two adjacent miner-

---

4. An exception to this statement is the use of dispersion staining methods, as described by Laskowski, Scotford, and Laskowski (1979), Laskowski and Scotford (1980), and Wilcox (1984).

als, or between a mineral and the mounting medium. It is usual to learn to recognize many of the common rock-forming minerals by sight, so with this knowledge the indices of an unknown mineral can be compared with a mineral of known indices. The relief shown by a mineral also provides a rough estimate of its refractive index or indices.

Alternately, thin sections are often made without cover glasses. Here again, the refractive index of an unknown material can be compared to adjacent minerals or the surrounding mounting medium. But also in this case, precise index measurements can be obtained after carefully chipping out a small amount of the substance and examining it in immersion liquids. Another approach is to chip away some of the adjacent mounting medium and apply immersion liquids directly to this part of the thin section; the liquids can be changed as necessary until a match is obtained. In addition, an uncovered polished thin section permits examination of opaque materials by reflected light, (a subject dealt with in detail in other volumes, such as Ramdohr (1969)).

IDENTIFICATION OF ISOTROPIC MATERIALS

Before using the microscope, observe macroscopic characteristics such as associated minerals, geological environment, hardness, color, cleavage, solubility, or chemical reactivity. These are often sufficient to provide a strong clue for identification.

With the microscope, the first step is to verify the isotropic character of the mineral by observing it during stage rotation after crossing the nicols. Isotropic materials are black, whereas anisotropic materials show colors, as explained more fully on p. 35, and in Chapter 6. Next obtain at least an approximate index of refraction, and note the color of the substance. In addition, note the presence or absence of cleavage, its quality when present, as well as angular relationships as seen in cross section. For example, cleavage serves to clearly distinguish periclase (MgO), with its perfect {100} cleavage, from spinel ((Mg,Fe)Al_2O_4), with its imperfect {111} parting. In the unlikely event that the above data are not sufficient for identification, it will be necessary to obtain a precise index of refraction, or resort to alternate techniques.

ADDITIONAL READINGS

Emmons, R. C., and R. N. Gates. 1948. The use of Becke line colors in refractive index determinations. *American Mineralogist 33*, 612–618.

Fisher, D. J. 1958. Refractometer perils. *American Mineralogist 43*, 777–780.

Fleischer, M., R. E. Wilcox, and J. J. Matzko. 1984. *Microscopic Determination of the Nonopaque Minerals.* U.S. Geological Survey Bulletin 1627, 3–9.

Jones, N. W., and F. D. Bloss. 1980. *Laboratory Manual for Optical Mineralogy.* Minneapolis: Burgess, 2-1 through 5-3.

Saylor, C. P. 1935. Accuracy of microscopical methods for determining refractive index by immersion. *Journal of Research National Bureau of Standards 15*, 277–294.

Wahlstrom, E. E. 1979. *Optical Crystallography,* 5th ed. New York: John Wiley & Sons, 107–149.

4

Uniaxial Materials and Light, I

Get your facts first, then you can distort 'em as you please.
MARK TWAIN

Anisotropic crystalline materials, unlike isotropic materials, have more than one characteristic index of refraction; each of these indices is directionally dependent. Measured indices of refraction of randomly oriented anisotropic materials are either the characteristic values or values intermediate to them. Consequently, the orientation of the crystal must be known before meaningful measurements of n can be obtained.

UNIAXIAL MATERIALS AND POLARIZATION

Anisotropic materials can be subdivided into two groups on the basis of differences in their optical behavior. Tetragonal and hexagonal materials are classified as uniaxial, whereas orthorhombic, monoclinic, and triclinic materials are biaxial. The terminology arises from the fact that all tetragonal and hexagonal crystals have a single direction along which the crystal exhibits isotropic behavior, that is, a light ray transmitted along this direction enters and leaves the crystal with its type and direction of polarization unchanged. This single direction, called the *optic axis*, is parallel to the c crystallographic axis in tetragonal and hexagonal materials. As there is only a single optic axis, such materials are called *uniaxial*. Orthorhombic, monoclinic, and triclinic materials have two such directions and are called *biaxial*.

The reason for the uniqueness of the c axis direction in uniaxial crystals becomes clear when we consider the atomic arrangement of uniaxial crystals perpendicular to the c axis. In tetragonal crystals (Fig. 4-1), the atomic arrangement is identical parallel to both the a_1 and the a_2 axial directions. Consequently, the dipole moments of atoms in both the a_1 and a_2 directions are identical. Effectively, the (001) plane "looks" isometric to a light ray traveling parallel to the c axis. Since mutually perpendicular directions of maximum and minimum polarization do not exist, the type of polarization possessed by the light ray traveling along the c axis is unchanged. Hexagonal materials are the same except that they have three identical a axes, not two.

However, a light ray transmitted along any direction other than the c axis within a uniaxial crystal traverses atomic packing arrangements that are distinctly anisotropic. When considering a tetragonal or hexagonal prismatic surface, it may be seen that packing along the c axis direction is different from that along the a axis direction (Fig. 4-2). As a consequence, a nonpolarized light ray traveling perpendicular to the c axis, or any other randomly chosen direction within the crystal, becomes doubly polarized.

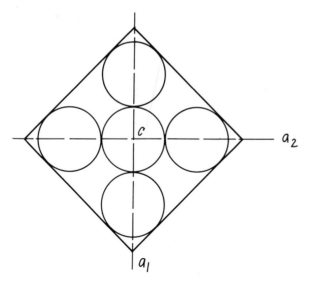

Figure 4-1
A possible atomic arrangement of a tetragonal crystal viewed along the c axis. With only this view available, the crystal could be interpreted as being isometric.

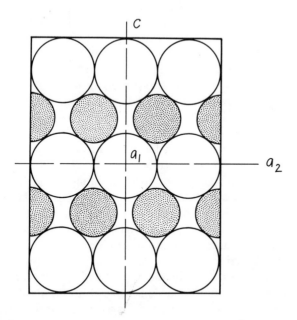

Figure 4-2
The crystal from Figure 4-1, viewed normal to the c axis. The atomic arrangement parallel to the a axis is different from that parallel to the c axis.

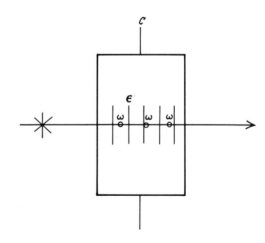

Figure 4-3
A beam of nonpolarized light passes through a uniaxial crystal at right angles to the c axis. Upon entering the crystal, the light becomes doubly polarized. One vibration plane contains the c axis and is called a principal plane. The vibration direction within this plane is shown by short vertical lines (parallel to the c axis); the index of refraction of this vibration is called epsilon (ε). A second vibration plane is normal to the c axis (and the page). The vibration direction within this plane is normal to the c axis and is shown by small circles; the index of refraction of this vibration is called omega (ω).

Let us consider first a ray of nonpolarized light incident to a uniaxial crystal. Subsequently, uniaxial grains will be considered with incident plane-polarized light—the arrangement in a polarizing microscope.

In Figure 4-3 a nonpolarized light ray is shown incident to a uniaxial crystal; the propagation direction of the incident ray is perpendicular to the crystal's c axis. Any plane containing the c axis (such as that in Fig. 4-2) is anisotropic; consequently, upon entering the crystal, the nonpolarized light is constrained to vibrate in two mutually perpendicular planes—that is, it becomes doubly polarized light. One of the vibration planes contains the c axis of the crystal and is called a *principal plane*. In Figure 4-3, it is a plane parallel to the page. The index of refraction of

28 Chapter 4: Uniaxial Materials and Light, I

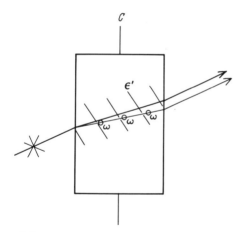

Figure 4-4
A beam of nonpolarized light passes through a uniaxial crystal at an oblique angle to the c axis; upon entering the crystal, the light becomes doubly polarized. The vibration direction within the principal plane, shown by short inclined lines, is inclined to the c axis; the index of refraction is ε' (an index between ε and ω in value). A second vibration plane is normal to the c axis (and the page). The vibration direction within this plane is normal to the c axis, and is shown by small circles; the index of refraction of this vibration is ω. The ray directions for ω and ε' diverge because each is refracted differently.

light vibrating in this plane is called epsilon (ε); the vibration direction within this plane is shown as vertical lines. The second vibration plane is perpendicular to the c axis (and the page). The vibration direction within this plane is shown as small circles; the index of refraction of light vibrating in this plane is called omega (ω). These two vibration planes along which the crystal permits light to vibrate are called the *permissible directions* or *privileged directions*. Other privileged directions are present in the crystal; they are a function of the angle between the incident beam and the c axis.

Uniaxial crystals can be subdivided according to the relative values of the indices of refraction ε and ω. If ε is greater than ω (if $\varepsilon - \omega > 0$), the crystal is *uniaxial positive*; if ε is less than ω ($\varepsilon - \omega < 0$), the crystal is *uniaxial negative*. The numerical difference between ε and ω, or between the maximum and minimum indices of refraction of any crystalline substance, is called the *birefringence*.

In Figure 4-4, a ray of nonpolarized light is shown entering a uniaxial crystal at an oblique angle to the c axis. Again, two privileged planes of vibration are produced; one is perpendicular to the principal plane (perpendicular to the page). The vibration direction within this plane is perpendicular to the c axis, as was true in the previous example (Fig. 4-3). As the vibration direction is maintained in spite of changing the orientation of the incident beam, we can presume that the index of refraction within this vibration plane is likewise unchanged from that in Figure 4-3. This is correct. In both examples, the index of refraction is ω.

If we examine the other privileged vibration plane (parallel to the page), it is obvious that this is a principal plane (being parallel to the c axis); however, in contrast to the earlier example in Figure 4-3, the vibration direction within this plane is not parallel to the c axis. Changing the direction of the incident beam has changed the vibration direction within the principal plane. The index of refraction produced is no longer ε; it is instead a value that lies between ε and ω, called ε' or "epsilon-prime."

If we suppose that the incident beam had entered the crystal at a different angle from the one chosen in Figure 4-4, it would follow that the vibration direction within the principal plane would be at a different angle to the c axis, and a different ε' would be produced. In fact, the value of ε' is a function of the orientation of the vibration direction within the principal plane. This relationship of ε' to the inclination θ of the vibration direction to the C axis is illustrated in Figure 4-5. A uniaxial negative mineral is assumed, in which ω is 1.72 and ε is 1.60. Vibration direction A, in the principal plane and parallel to the c axis ($\theta = 0°$), has an index of refraction of 1.60—the value of ε. Vibration direction B, at 30° to the c

axis, has a higher index of refraction, in this case 1.627. Vibration direction C, at 45° to the *c* axis, yields a still higher value, as does vibration direction D. The maximum value, obtained for any vibration direction in a uniaxial negative crystal, is equal to the value of ω, namely 1.72. When the vibration direction is perpendicular to the *c* axis ($\theta = 90°$), ε' and ω become identical and are simply called ω. When the two vibration directions within the crystal have identical indices, this means that the substance is behaving isotropically. Hence any ray passing along the *c* axis shows isotropic behavior, and its type of polarization is not affected by the crystal.

If, in our example, we had chosen a uniaxial positive material (where $\varepsilon > \omega$), values of ε' would again change as a function of the angle between the vibration direction and the *c* axis. But here ε' values would decrease toward the ω value as the vibration direction became more nearly normal to the *c* axis.

The changes in ε' (for both positive and negative crystals) do not occur in a one-to-one relationship with the angle θ between the vibration direction and the *c* axis; if the vibration direction is at 45° to the *c* axis ($\theta = 45°$), for example, the value of ε' is not exactly halfway between ε and ω. Instead the value of ε' follows an equation that describes the radii of an ellipse:

$$\varepsilon' = \frac{\omega}{\sqrt{1 + \left(\frac{\omega^2}{\varepsilon^2} - 1\right)\cos^2\theta}}$$

RAY VELOCITY SURFACES

The velocity of light waves associated with the ε' indices, by definition, likewise varies as a function of direction. This variation can be visualized by considering *ray velocity surfaces*. First imagine that a light is turned on in the center of an isotropic material. Light waves move out at constant velocity in all directions from the source. If the

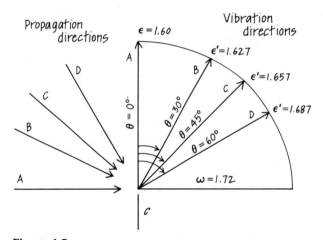

Figure 4-5
An example of the difference in index of refraction ε' as a function of the change in *vibration direction*. The vibration direction parallel to the *c* axis yields the index ε (here $\varepsilon = 1.60$). As the vibration direction approaches perpendicularity with the *c* axis, the value of ε' changes so as to approach that of ω (here ω = 1.72).

light waves were stopped at any given instant while still within the grain, a locus of points showing the distance of travel in all directions could be visualized as a solid object; such an object is called a ray velocity surface. As the velocity from the point source is equal for all directions within an isotropic material, the ray velocity surface is a sphere. (Note that there is a unique ray velocity surface for each wavelength of light.)

Within uniaxial materials, different velocities are associated with both the $\varepsilon(\varepsilon')$ and ω indices. Consequently, two ray velocity surfaces are generated for any particular wavelength. As the velocities of the ω vibration directions do not vary as a function of direction of propagation, the ω ray velocity surface is a sphere. The ε vibration directions vary in velocity as a function of direction, and their ray velocity surface is either a prolate (elongate) or oblate (flattened) spheroid (an ellipsoid of revolution).

Consider the ε ray velocity surface for a uniaxial positive mineral (where $\varepsilon > \omega$). When the ε

30 Chapter 4: Uniaxial Materials and Light, I

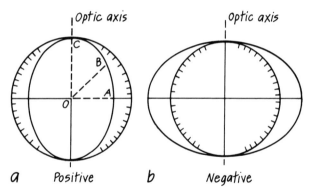

Figure 4-6
Ray velocity surfaces of ε and ω viewed in cross section. The rays that produce ω vibrations yield spherical ray velocity surfaces (hatched). Rays that produce ε vibrations yield prolate (A) or oblate (B) spheroids.

ray travels perpendicular to the c axis its velocity is minimal (line OA, Fig. 4-6A), as it yields the ε value (and index of refraction is inversely proportional to velocity). A ray traveling at an oblique angle to the c axis yields an intermediate ε' value and a velocity between that of ε and ω (line OB, Fig. 4-6A). A ray traveling parallel to the c axis (line OC) and vibrating normal to c has an ε' index that corresponds to ω; consequently, the ray velocity surfaces for both ε and ω coincide at the c axis. A uniaxial positive mineral has an ε ray velocity surface that is a prolate spheroid (Fig. 4-6A), whereas the ε ray velocity surface of a uniaxial negative mineral is an oblate spheroid (Fig. 4-6B).

Snell's law, discussed earlier, does not take into account that uniaxial (and biaxial) crystals possess spheroidal ray velocity surfaces, hence it cannot be used to determine ray paths with ε-type light waves. Snell's law does permit determination of the *wave normal* direction—the direction normal to the vibration direction. The deviation from Snell's law of the ε ray path was regarded as extraordinary, and hence the ray direction associated with ε-type vibrations are often called extraordinary (or E) rays, in contrast to the ordinary behavior of rays associated with ω vibrations (often called O rays).

The behavior of the ε vibrations has been called extraordinary because of the fact that a light ray at perpendicular incidence to a crystal surface may produce an E-ray path whose angle of refraction is other than zero. Snell's Law predicts a zero angle of refraction. This discrepancy is explained by the spheroidal ray velocity surfaces for ε-type vibrations

Consider the uniaxial crystal with a pyramidal surface shown in Figure 4-7A; light from the upper left has perpendicular incidence. The direction of the refracted beam can be determined by means of a construction devised by Christian Huygens in the seventeenth century. Huygens stated that the front of each light wave at the crystal edge can be regarded as the source of a secondary wave that spreads out in all directions from that point. The direction of the refracted rays is obtained by finding a new wave front common to all of the possible incident rays. In Figure 4-7A, the beam of light is composed of a multitude of parallel rays such as A, B, and C. The front of these waves (the *wave front*) is parallel to the surface of the crystal during the approach. When the waves meet the surface, each point of contact (such as A', B', and C') acts as a secondary source of waves that can move into the crystal in any direction. The possible travel distance of these secondary waves at some later time is shown by ray velocity surfaces. As Figure 4-7A considers the ω vibrations, the ray velocity surfaces are hemispheres. The line DE, drawn tangent to these ray velocity surfaces, represents the wave front within the crystal. In order to find the direction of travel of the rays within the crystal, it is necessary to draw lines from the source of the secondary waves (at A', B', and C') to the intersection points (A", B", and C") of the ray velocity surfaces with the wave front DE. The lines A'A", B'B", and C'C" represent the direction of travel of the rays within the crystal. It can be seen that the angle of incidence and the angle of refraction are both zero and that the rays cross the interface with no deviation; thus Snell's law is satisfied (which no doubt pleases his descendants).

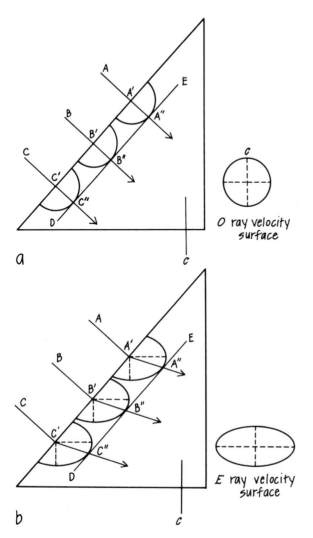

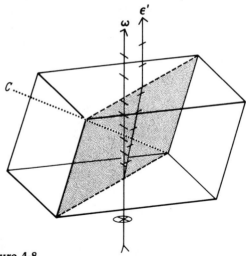

Figure 4-8
A ray of nonpolarized light that enters a calcite cleavage fragment with perpendicular incidence to a rhombohedral surface becomes doubly polarized. The ω vibrations travel through the crystal without refraction. The ε' vibrations are refracted when crossing the interface and follow a different path. An observer viewing the crystal from above sees two spots of light, each of which is plane-polarized at right angles to the other. A vertical principal plane is shaded.

Figure 4-7
Huygens's construction showing the use of ray velocity surfaces for rays entering a crystal with perpendicular incidence and at an oblique angle to the c axis. The construction is shown in (A) for ω vibrations and in (B) for ε and ε' vibrations. See text for explanation.

Consider the result of a Huygens construction involving the ε vibrations (Fig. 4-7B); here the ray velocity surfaces are elliptical. This causes the refracted beams A'A", B'B", and C'C" to have an angle of refraction that is not zero. The ray direction and the wave front are not perpendicular within the crystal. In addition, the vibration direction (parallel to the wave front) is not normal to the ray direction. Snell's law does not predict this, as it applies only to a spherical ray velocity surface.

Deviation from Snell's law may be seen easily in Iceland spar—clear cleavage fragments of calcite (Fig. 4-8). A beam of light with perpendicular incidence to a rhombohedral cleavage surface separates into two different types of vibrations, each of which has a different path. The ω vibrations pass straight through the crystal with no deviation; this occurs because the *O*-ray velocity surface is a sphere. The ε' vibrations, whose ray velocity surface is an oblate spheroid, follow a different path. This is the famous "double refraction" of calcite. In fact, all uniaxial and biaxial minerals show double refraction, but the effect in most is not as obvious as that shown by calcite, first because large clear crystals of minerals other than calcite are uncommon, and second because the index difference between ε and ω (the birefringence) is usually not as great for most minerals as for the carbonate minerals.

The difference in refractive angles between ω and ε vibrations will be ignored in most figures in this book. Detailed explanations of the phenomenon are given by Bloss (1961, pp. 65–90).

THE OPTICAL INDICATRIX

A slightly different way of describing the indices of refraction of uniaxial materials is in terms of the *optical indicatrix*, a three-dimensional geometric representation of the indices of refraction of a substance. The radii of the indicatrix are proportional in length to the refractive indices, in the direction of their vibration, for all waves originating at a point source. In some ways an indicatrix can be regarded as a reciprocal ray velocity surface that represents both ω and ε indices (as $n = 1/V$). As with ray velocity surfaces, the indicatrix is a monochromatic representation surface.

Positive and negative uniaxial indicatrices are shown in Figure 4-9A. The vertical direction is parallel to the optic (*c*) axis of the crystal. Perpen-

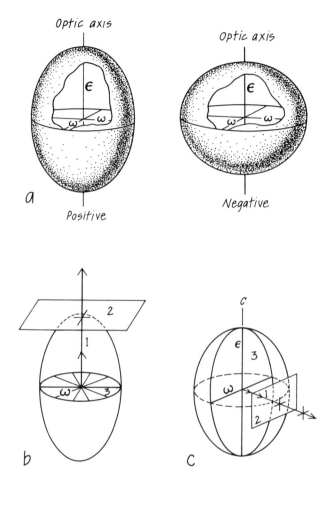

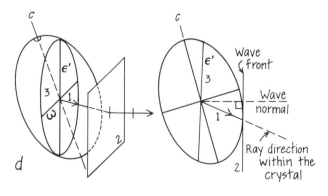

Figure 4-9
(A) The uniaxial indicatrix. The optic axis is parallel to the *c* crystallographic axis. The length of the radius parallel to the optic axis is proportional to the value of the index of refraction ε. Perpendicular to the optic axis is a circular section whose radius corresponds to the index ω. Intermediate radii correspond in length to various values of ε'. The uniaxial indicatrix is either a prolate spheroid (for optically positive materials) or an oblate spheroid (for optically negative materials). (B–D) Three-step procedure for finding the vibration directions and indices of refraction for a particular ray direction. See text for explanation.

dicular to the *c* axis, the indicatrix contains a *circular section*. The vertical radius corresponds in length to the ε index, and the radii of the circular section correspond in length to the ω index. Intermediate radii correspond in length to various values of ε'. The shape of the positive uniaxial indicatrix is a prolate spheroid; that of the negative uniaxial indicatrix is an oblate spheroid. As an aid to memory, the tall positive indicatrix can be related to the vertical stroke of a plus sign and the squat shape of the negative indicatrix to a minus sign.

The remaining parts of Figure 4-9 show one use of the indicatrix: the determination of indices of refraction for various ray directions within a uniaxial (positive) crystal. This can be demonstrated in a three-step sequence (numbered in the figure): (1) show the direction of the light ray from the center of the indicatrix to the indicatrix surface; (2) where the ray intersects the indicatrix surface, indicate a plane tangent to the surface, representing the wave front of the ray; (3) draw a plane parallel to this wave front through the center of the indicatrix; this plane is an elliptical or circular section whose maximum and minimum radii represent the indices of refraction of the ray.

In Figure 4-9B, the ray (1) travels parallel to the *c* axis, and the resulting plane within the indicatrix (3) corresponds to the circular section. Any vibration within this wave front must have the index ω, as all radii of the circular section are equal in length to ω.

In Figure 4-9C, the ray (1) travels perpendicular to the *c* axis, the tangent (2) is parallel to the *c* axis, and the wave front (3) within the indicatrix is an ellipse whose longest radius corresponds in length to ε; the shortest radius falls within the circular section and has the index ω.

Part D shows a ray (1) at an oblique angle to the *c* axis. The wave front (2) tangent to the indicatrix surface is *not* perpendicular to the ray direction. (This was also true in the examples shown earlier in Figures 4-4, 4-5, and 4-7B.) The wave front (3) within the indicatrix is an ellipse whose largest radius has the length ε' and whose shortest radius again falls within the circular section and has the index ω. These two radii represent the two vibration directions of the ray, and their lengths represent the two indices produced by the ray.

Any ray that is neither parallel nor perpendicular to the *c* axis of a uniaxial crystal has a wave front that is not perpendicular to the ray direction. A line drawn perpendicular to the wave front is called a *wave normal*, as it is normal to the ray's vibration direction. Snell's law permits a calculation of the wave normal direction, but does not permit a calculation of the ray direction if the two are not parallel.

UNIAXIAL CRYSTALS VIEWED WITH A MICROSCOPE

Under the microscope, uniaxial crystals are subjected to plane-polarized light from the lower polarizer. This light is generally restricted to an east-west (right-left) vibration direction relative to the observer (Fig. 4-10). Most lower polarizers can be rotated; a detent or zero point to which the lower polarizer can be set is usually present. In many older instruments, the zero setting puts the permissible vibration direction of the lower polarizer in a north-south orientation, but since 1972, by international agreement, east-west has been the standard. This book will use the abbreviations NS and EW.

The lower polarizer setting of an unfamiliar microscope must be determined before using the instrument with anisotropic crystals. The orientation is usually determined with the aid of a mineral that shows strong differential absorption. Biotite, for example, strongly absorbs light vibrating parallel to its direction of cleavage (as seen in thin section). If the lower polarizer is oriented EW, biotite will appear darkest when its direction of elongation is EW; if the lower polarizer is oriented NS, biotite will appear darkest when its direction of elongation is NS.

34   Chapter 4: Uniaxial Materials and Light, I

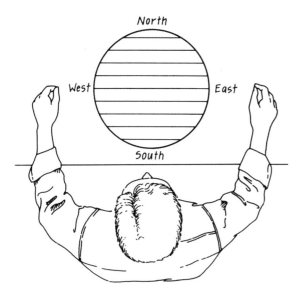

Figure 4-10
The conventional reference directions that are used with the microscope, as related to the microscopist.

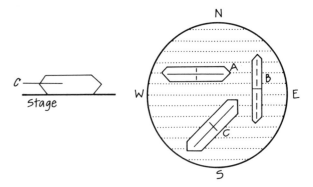

Figure 4-11
Uniaxial crystals as viewed in the microscope. The crystals are elongated on their c axis, which is parallel to the stage. Solid and dashed lines within the crystals indicate the privileged vibration directions. Dotted lines represent EW plane-polarized light from the lower polarizer.

The NS versus EW setting of the lower polarizer affects the measurement of indices of refraction. The vibration direction whose index of refraction is to be measured must be oriented parallel to the privileged direction of the lower polarizer. We will assume henceforth that our instrument is set so that all incident light coming to the specimen is vibrating in the EW plane.

Three optically different orientations are possible for a uniaxial crystal: the c axis is parallel to the stage; the c axis is inclined to the stage; the c axis is perpendicular to the stage. These will be considered in turn.

Optic Axis Parallel to the Stage

Let us take three uniaxial crystals as viewed in the microscope, with the c axes of each parallel to the stage. Each crystal is shown in Figure 4-11 elongated parallel to the c axis, whereas normally fragment shapes are dependent upon cleavage characteristics. The lower polarizer is in place, as always, and the EW plane-polarized light is de-

picted with dotted lines. Lines within the crystals indicate the orientation of the privileged vibration planes; note again that one of these is always a principal plane (parallel to the c axis) while the other is perpendicular to it.

In addition to all of this, let us assume that the upper polarizer has been inserted. This situation is usually described as *crossed nicols*, as the single privileged direction of the upper polarizer is normally oriented at right angles to that of the lower polarizer. ("Uncrossed nicols" is sometimes used to mean that only the lower polarizer is in place, but the more logical expression is "plane-polarized light.") The upper polarizer will only transmit light that is vibrating in or has a component in the NS plane. Inserting the upper polarizer eliminates all light that has not gone through the grains. If no crystals were present, the field of view would appear black.

What is the effect of crossed nicols on the crystals? Crystal A is oriented so that its principal plane is parallel to the privileged direction of the lower polarizer. Therefore the plane-polarized light travels through the crystal and continues to be purely plane-polarized. That is why the NS privileged direction is depicted with a dashed line. This plane-polarized light leaves the top of the grain and is eliminated by the upper polarizer, hence the crystal appears black against a black background; that is to say, nothing appears at all. It is said to be *at extinction*.

Consider crystal B. Again one of the privileged directions in the crystal (the other one this time) is parallel to the incident plane-polarized light. Again the plane-polarized light is transmitted through the crystal without becoming doubly polarized; again this light is eliminated by the upper polarizer. This crystal is also at extinction.

The third crystal, C, is oriented so that neither of its privileged directions is parallel to the incident plane-polarized light. Passing through this crystal, the plane-polarized light breaks into two vibrational components (Fig. 4-12), one in the principal plane of the crystal and the other in the plane perpendicular to the principal plane. These two vibrations travel separately through the crystal, each with its own velocity and wavelength. Upon leaving the crystal, the vibrational constraints are removed, and the two vibrations combine into a single vibration. The result is most commonly elliptically polarized light (but may be circularly polarized or plane-polarized). As elliptically polarized light has vibrational components in all directions, much of this light gets through the upper polarizer and reaches the microscopist's eye (or eyes). The grain now appears against the black background, showing one or more colors. These colors are called *interference colors* because they originate from the interference of light from the two planes of vibration. They are not related to the normal color of the mineral.

What have we established by all of this? Quite a lot. We have determined that when either of the two privileged directions in a grain is parallel to the incident plane-polarized light, the crystal is at extinction; when neither is parallel, the crystal transmits light. Therefore, we have a method of finding the privileged directions. All that is necessary is to view a crystal between crossed nicols and rotate the stage until the crystal becomes black. When this occurs, one of the privileged di-

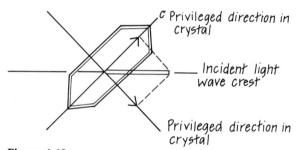

Figure 4-12
Crystal C (from Fig. 4-11) in diagonal orientation on the stage. The plane-polarized light from the lower polarizer (double lines) is resolved into two vibration planes within the crystal. This is done graphically by erecting lines (dashed) from the incident wave crest perpendicular to the two privileged directions within the crystal. The initial wave amplitudes within the crystal are shown by arrows.

rections of the crystal matches the EW privileged direction of the lower polarizer. Continued rotation of the stage will produce another extinction (at 90° from the first); now the other privileged direction is EW.

In addition to locating the privileged directions in uniaxial materials, this approach serves to distinguish anisotropic crystalline materials from isotropic substances. In most orientations, anisotropic materials display colors (or shades of gray) and extinctions during rotation with crossed nicols, whereas isotropic materials are incapable of producing doubly polarized light, and consequently remain in constant extinction.

Optic Axis Inclined to the Stage

Consider the second type of orientation, with the c axis inclined to the stage. Here we will use several cleavage fragments of the uniaxial mineral calcite (Fig. 4-13). The perfect $\{10\bar{1}1\}$ rhombohedral cleavage of calcite orients cleavage fragments such that the c axis is neither parallel nor perpendicular to the stage. (In plan view in the figure, this direction is labeled c' to indicate that the c axis is at an oblique angle to the stage.)

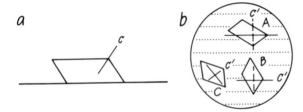

Figure 4-13
(A) Vertical section of a cleavage fragment lying on a rhombohedral surface. (B) Rhombohedral cleavage fragments as viewed in the microscope. Solid and dashed lines within the fragments indicate the privileged vibration directions.

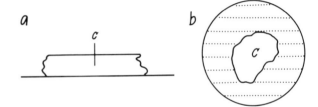

Figure 4-14
(A) Vertical section of a cleavage fragment lying on a basal pinacoid surface. (B) Basal pinacoid cleavage fragment as viewed in the microscope. The c axis is perpendicular to the stage and the crystal exhibits isotropic behavior. Grain edges are irregular.

Again, the privileged vibration directions are indicated within the fragments by solid and dashed lines; one direction is a principal plane and the other is perpendicular to it. The situation is very much like that presented in Figure 4-11. Grains A and B have a privileged direction parallel to the incident plane-polarized light; this light passes through the crystals without becoming doubly polarized and is eliminated by the upper polarizer. Both grains are at extinction. Grain C has neither privileged direction parallel to the incident plane-polarized light; thus the light transmitted through C is doubly polarized. Recombination at the top of the grain generally yields elliptically polarized light, part of which passes through the upper polarizer to be seen as interference colors.

Optic Axis Perpendicular to the Stage

The third type of orientation places the c axis perpendicular to the stage (Fig. 4-14). We can imagine the grain to be a fragment of the uniaxial mineral brucite, lying on a perfect (001) cleavage surface. The ray path of the incident light is parallel to the c axis of the crystal. As the c axis is also the optic axis, this means that the crystal shows isotropic behavior for this orientation; that is, the incident ray enters and leaves the crystal plane-polarized EW. This light is eliminated by the NS-oriented upper polarizer, the crystal is at extinction, and as the c axis remains vertical, rotation of the stage does not change this situation. A crystal such as this is superficially indistinguishable from an isotropic material. Fortunately, other techniques, such as interference figures (Chapter 6), can reveal the difference between isotropic and anisotropic materials.

MEASURING INDICES OF REFRACTION OF UNIAXIAL MATERIALS

In determining the indices of refraction of uniaxial materials, interference figures are commonly used to verify grain orientation. As this technique has not yet been treated, we will go through an exercise to show how indices can be determined without interference figures. This will also help reinforce some of the preceding concepts.

Assume that an unknown uniaxial mineral has been crushed and sieved to standard size. The mineral has no cleavage, so grains may lie in any orientation. Figure 4-15 shows a group of fragments in the field of view of the microscope. Grains A, B, and C are the most common type; each is oriented with its c axis (shown as c') at an oblique angle to the plane of the stage. With the nicols crossed, such grains show interference colors and extinctions at 90° intervals of stage rota-

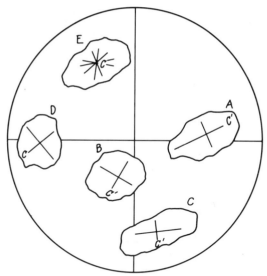

Figure 4-15
A variety of grains in different orientations. Permissible vibration directions are indicated by solid lines within the grains. Grains A, B, and C are oriented such that the c axis is oblique to the stage (indicated as c'). The c axis of grain D is parallel to the stage. The c axis of grain E is perpendicular to the stage; permissible vibration directions for grain E are in all planes that include the c axis.

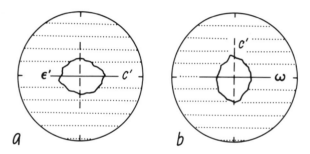

Figure 4-16
A grain (such as A, B, or C in Fig. 4-15) whose c axis is oblique to the stage. (A) The grain is in an extinction position, such that the trace of the c axis (indicated as c') is oriented EW (parallel to the permissible vibration direction of the lower polarizer). When the nicols are uncrossed, the Becke test permits a measurement of ε' in this orientation. (B) The same grain is rotated 90° into another extinction position. With uncrossed nicols in this orientation, the Becke test permits a measurement of the index ω.

tion, each winking out at a different point as the stage turns. A less common grain orientation is D, whose c axis is parallel to the stage; interference colors are present, and the grain shows extinctions at 90° intervals during rotation of the stage. Grain D therefore resembles grains A, B, and C. A very uncommon orientation is grain E, whose c axis is perpendicular to the stage. This grain remains in constant extinction during stage rotation.

What information can we extract from these grains? Grains A, B, and C can each be rotated into an extinction position. In Figure 4-16A a grain of this type has been rotated into an extinction position, such that the principal plane is EW. Vibrations within the principal plane have an index of refraction ε' (see Fig. 4-4), the value of which varies as a function of the angle between the crystal's c axis and the microscope stage. Removing the upper polarizer from the system (or "uncrossing the nicols") permits a Becke test to be made for this index of refraction. After this is done, the upper nicol can be reinserted and the grain rotated 90° to the next extinction position. The vibration transmitted by the grain is now perpendicular to the principal plane—that is, the ω index can be measured with a Becke test after removing the upper polarizer (Fig. 4-16B).

Now, consider a grain whose c axis is horizontal (such as grain D in Fig. 4-15). The grain has two different extinction positions (see Fig. 4-17); one permits a Becke test of the ε index, the other a Becke test of ω. Grain D, again, is superficially indistinguishable from grains A, B, and C.

Grain E has its c axis vertical (Fig. 4-18). The light ray path is parallel to c, and the vibration direction is perpendicular to c; this grain permits a Becke test of the ω vibration (Fig. 4-5). Rotation of the stage does not change this orientation.

These relationships are simple to summarize. It can be seen that a uniaxial grain in *any* orientation will, when rotated into the proper position, per-

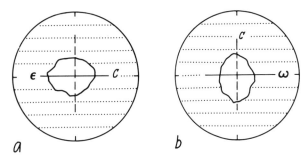
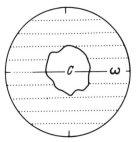

Figure 4-17
A grain (such as D in Fig. 4-15) whose c axis is parallel to the stage. (A) The grain is in an extinction position; when the nicols are uncrossed in this orientation, the Becke test permits a measurement of the index ε. (B) The same grain is rotated 90° into another extinction position. With uncrossed nicols, the Becke test permits a measurement of the index ω.

Figure 4-18
A grain (such as E in Fig. 4-15) whose c axis is vertical. This grain remains at constant extinction during rotation of the stage. With uncrossed nicols, the Becke test permits a measurement of the index ω.

mit a Becke test measurement of ω. However, measurement of ε is possible only when the grain has its c axis parallel to the stage of the microscope. Other orientations yield either values of ε' or of only ω. This explains why most determinative tables are based upon ω values rather than ε.

How do we determine ε and ω? Assuming random grain orientation and no knowledge of the indices of refraction of the mineral (let us say that ω is 1.635 and ε is 1.682 in this example), a mount is prepared, and a randomly chosen intermediate immersion liquid is used—for example, 1.57. Typically both ω and ε of the specimen are either above or below the index of the chosen immersion liquid, and all of the grains will show some relief. Therefore, it is useless at this time to seek out particular grain orientations. After making sure that the upper polarizer is removed and the substage diaphragm is partially closed, perform the Becke test on a number of grains. In this case the Becke lines move into all of the grains, which indicates that a second mount must be made with a higher index immersion liquid. The process of making grain mounts continues with ever-decreasing grain relief. A point will be reached when the chosen immersion liquid falls between the values of ε and ω—for example, an immersion liquid with n of 1.66. This situation is easily recognized; the Becke test shows an inward movement of the Becke line on some grains and an outward movement on others. This is an appropriate time to determine the optic sign.

The thing to do now is to find a grain that has a vertical or near-vertical c axis, as such a grain will permit a measurement of *only* ω. A search will reveal a few grains that remain in complete or almost complete extinction (perhaps becoming some dark shade of gray) during stage rotation. A Becke test on such grains will reveal either ω or a value of ε' that is very close to it.

Assuming an immersion liquid of $n = 1.66$, a Becke test reveals that ω is less than 1.66. It follows that if ω is less than 1.66, and some of the other grains on the slide show higher indices, the higher indices must be either ε or ε'. The mineral is, therefore, uniaxial positive. If ω had turned out to be greater than 1.66, with other grains present having lower indices, the mineral would be negative.

Knowing the optic sign of the mineral, it is now necessary to make separate mounts to determine each index. Start first with ω. We know now that $\omega < \varepsilon$. Every grain will permit a Becke test of ω.

Choose an immersion liquid less than 1.66, the particular choice being based on the relief. Suppose you had chosen 1.63. Any grain on the slide can be rotated into its two extinction positions and the two indices compared by a Becke line test. As the mineral is uniaxial positive, the position that yields the lower index must furnish an estimate of ω: the other position yields either ε' or ε. As the lower index turns out to be slightly higher than 1.63, we now know that ω falls between 1.63 and 1.66. Additional mounts, with intermediate liquids, will reveal the true value of 1.635.

The determination of ε is somewhat more involved, as vibrations in the principal plane appear as various values of ε', as a function of grain orientation. We have established that ε is greater than 1.66, so mounts are made with successively higher index immersion liquids. When each mount is made, grains should be chosen that have the following two characteristics: (1) they are of average or smaller than average size and (2) they possess a greater than average number of bands of interference colors (these are the colors seen with crossed nicols). If none of the grains have color bands but are white to gray, choose grains that show the least amount of gray.

As will be explained in Chapter 6, this approach gives moderate assurance that the c axis of the grain is parallel or near parallel to the stage, the position that permits a determination of ε or a value of ε' that is close to it. With successive mounts in different immersion liquids, a point will be reached where no grain can be found that has a higher index of refraction than the liquid. The true value of ε is probably located between this liquid and the next lower index liquid in which ε or ε' has been revealed to be higher.[1]

With this fairly crude approach, we cannot be sure that we have determined the actual ε value, but chances are that we have come close. When identifying the mineral in determinative tables (such as those in Volume 2 of this book), keep in mind that the value determined for ε might be, in fact, a close value of ε'. It can be assumed, however, that the determined value of ω is correct.

ADDITIONAL READINGS

Bloss, F. D. 1961. *An Introduction to the Methods of Optical Crystallography.* New York: Holt, Rinehart and Winston, 65–91, 235–236.

Hallimond, A. F. 1970. *The Polarizing Microscope,* 3d ed. York, U.K.: Vickers Instruments, 76–88, 236–250.

Hartshorne, N. H., and A. Stuart. 1969. *Practical Optical Crystallography,* 2d ed. New York: American Elsevier, 154–198.

Johannsen, A. 1914. *Manual of Petrographic Methods.* New York: McGraw-Hill (reprinted by Dover Publications), 61–90.

Loupekine, I. S. 1947. Graphical derivation of refractive index ε for the trigonal carbonates. *American Mineralogist* 32, 502–507.

Phillips, W. R. 1971. *Mineral Optics.* San Francisco: W. H. Freeman, 75–88.

Wahlstrom, E. E. *Optical Crystallography,* 5th ed. New York: John Wiley & Sons, 205–241.

---

1. It has been assumed in this example that the mineral does not have any cleavage. The presence of cleavage will result in preferred orientation of grains, which might prevent a determination of ε by this technique.

5

Uniaxial Materials and Light, II

No knowledge of music is necessary. Merely place kazoo to lips and hum.
  ANONYMOUS KAZOO INSTRUCTIONS

This chapter will consider additional factors relating to the identification of uniaxial materials: the effect of cleavage, the nature of interference colors, the use of accessory plates, and the observation of pleochroism. Knowledge of each of these subjects will aid in determining optical parameters and will lead directly to the origin and use of interference figures, as outlined in Chapter 6. With knowledge of interference figures, any uniaxial material becomes fair game for identification by even the beginning microscopist.

CLEAVAGE AND FRAGMENT ORIENTATION

Cleavage is an exceedingly valuable property for identifying unknown uniaxial materials. Although much of this discussion of cleavage will be limited to crushed fragments, cleavage is also visible in thin sections of rocks as a series of parallel cracks. The relationship of type and perfection of cleavage to optical directions and crystal habit will be seen to be extremely important to mineral identification in both thin sections and crushed grains.

Certain crystalline materials are characterized by excellent cleavage; others show no cleavage. Some minerals may show *parting,* which results in breakage along planes of weakness caused by polysynthetic twinning or deformation. No quantitative method of describing quality of cleavage has been universally accepted; instead it is usual to describe cleavage of decreasing perfection as perfect, good, distinct, imperfect, and poor. As the terms good, distinct, and imperfect are difficult to remember and have no numerical basis of distinction, cleavage descriptions will be simplified here to three categories—perfect, good, and poor.

If a mineral possesses *perfect cleavage,* most grains in a crushed sample are affected and lie in the same orientation; interference colors are constant across the grains, as a result of a uniform thickness. In thin sections, almost every grain (when oriented in a favorable position) shows parallel cracks that indicate the presence of this cleavage.

A *good cleavage* is one that affects a significant number of grains. With crushed fragments, many of the grains may show a distinctive shape, a preferred orientation, and a definite angular relationship at the extinction position between the grain's cleavage edge and the cross hairs. In thin section, cleavage cracks are common but not ubiquitous.

A *poor cleavage* is one that has little effect on grain orientation of crushed fragments and that rarely shows up as parallel cracks in thin section. This type of cleavage may be overlooked until the final stage of mineral identification. A common situation is for the microscopist to reach the point

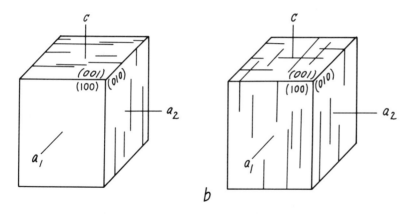

Figure 5-1
(A) Cleavage parallel to (100) is shown within a block diagram. (B) Cleavage parallel to {100} in a tetragonal crystal results in two differently oriented, symmetrically equivalent cleavages, one parallel to (100) and another parallel to (010).

Table 5-1
Uniaxial cleavage types related to grain orientation on the microscope stage

Tetragonal system	Hexagonal system	Orientation of *c* axis	Number of cleavage directions
Basal pinacoid, {001}	Basal pinacoid, {0001}	Perpendicular to the stage	1
Prismatic, {100}, {110}, {hk0}	Prismatic, {10$\bar{1}$0}, {11$\bar{2}$0}, {hk$\bar{\imath}$0}	Parallel to the stage	2 or more
Pyramidal, {h0l}, {11l}, {hkl}	Pyramidal, rhombohedral, or scalenohedral, {10$\bar{1}$l}, {11$\bar{2}$l}, {hk$\bar{\imath}$l}	Inclined to the stage	3 or more

of attempting to decide between two or three minerals with similar characteristics. The description of one might mention the presence of a poor cleavage, and such a cleavage might emerge upon reexamining the crushed fragments or thin section. Basing an identification on such a cleavage, however, is risky because a poor cleavage (although described in the literature) is commonly absent within particular specimens. Supplementary methods are often necessary in such cases.

The orientations of cleavage surfaces are described in terms of Miller indices. We will use two standard types of Miller index notations. A Miller index such as (100) refers to a single set of cleavage surfaces, all of which are parallel to the *faces* (100) and ($\bar{1}$00) (Fig. 5-1A). The Miller index {100} refers to all of the surfaces that constitute the *form* {100}. In the tetragonal system, {100} yields the four surfaces of a second-order prism: (100), ($\bar{1}$00), (010), and (0$\bar{1}$0). Thus, a tetragonal crystal described as having {100} cleavages will yield two directions of cleavage, one set of cleavage surfaces parallel to (100) and ($\bar{1}$00) and a second perpendicular set parallel to (010) and (0$\bar{1}$0) (Fig. 5-1B).

Cleavages in uniaxial minerals can be subdivided according to their angular relationship to the *c* axis (Table 5-1).

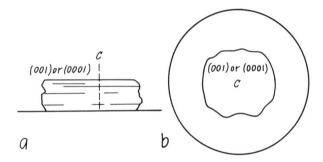

Figure 5-2
A basal cleavage—{001} for tetragonal materials, and {0001} for hexagonal materials. Symmetry requires only a single cleavage direction. (A) The single basal pinacoid cleavage, as viewed in vertical section. Grains on this cleavage have their c axis perpendicular to the stage. (B) Seen in plan view the single basal pinacoid cleavage results in a fragment of irregular outline. The grain remains at extinction during stage rotation.

Basal cleavages in both the tetragonal and hexagonal systems produce a single group of parallel planes because the basal pinacoid form produces two parallel surfaces such as (001) and (00$\bar{1}$), or (0001) and (000$\bar{1}$). When such cleavages are perfect (as with chlorite and brucite), almost all grains lie on surfaces parallel to the basal pinacoid; grain outlines as seen in the microscope are typically irregular because no other cleavages are present. The c axis is perpendicular to the stage (Fig. 5-2), so when these grains are viewed between crossed nicols during stage rotation, the fragments remain in extinction.

Some cleavages produce surfaces at acute angles to the c axis. In the tetragonal system, the dipyramidal cleavages {h0l} and {hhl} produce four different directions of breakage, and the {hkl} cleavage produces eight different directions (Fig. 5-3A). These cleavages are uncommon. In hexagonal materials, the {10$\bar{1}$l} cleavage may produce either three or six differently oriented cleavage surfaces (depending upon crystal class); the {11$\bar{2}$l} cleavage produces six; and the {hk$\bar{i}$l} cleavage produces six (scalenohedron) or twelve (dihexagonal dipyramid) directions (Fig. 5-3B). When grain fragments are viewed in the microscope, all of these different cleavage types look essentially the same (Fig. 5-3C): the microscopist's plan view is that of a parallelogram-shaped grain. The internal angles of the parallelogram vary as a function of cleavage angles and types. Fortunately, in most cases determining the angles between these cleavages is not critical for mineral identification. The main thing is to learn how to recognize that a pyramidal, rhombohedral, or scalenohedral cleavage is present.

The type of extinction is as important as the shape of the grain. As shown in Figures 4-12 and 5-3C, the c axis of each cleavage fragment is disposed symmetrically with respect to the various cleavage surfaces. Thus, when the fragment is at extinction, the cleavage edges are disposed symmetrically to the NS and EW cross hairs; that is, the cross hairs (or lines drawn parallel to them) bisect the acute and obtuse internal angles between the cleavages. Pyramidal, rhombohedral, or scalenohedral cleavage in uniaxial materials is indicated by *symmetrical extinction*.

Prismatic cleavages are parallel to the c axis. Tetragonal prismatic cleavages are {100}, {110}, or {hk0}. These result in either two or four directions of cleavage. Hexagonal prismatic cleavages are either {10$\bar{1}$0}, {11$\bar{2}$0}, or {hk$\bar{i}$0}; these yield either three or six sets of cleavage surfaces (Fig. 5-4A). All are parallel to the c axis of the fragment (as indicated by the zero in the last part of the Miller index). If only prismatic cleavage is present, at least two directions of breakage develop; this results in fragments that are elongated in the c axis direction and have their straight sides parallel to c.

Such cleavage fragments are easy to recognize; at their positions of extinction, the straight edges of the grain are parallel to the NS or EW cross hairs because the c axis direction corresponds to a principal vibration plane. This type of extinction is called *parallel extinction*.

After the grain shapes and types of extinction are recognized, it then becomes obvious which indices of refraction can be measured. Grains on

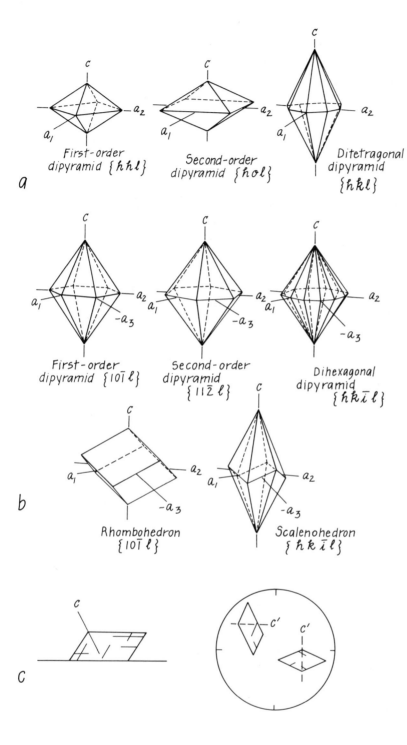

Figure 5-3
(A) Pyramidal cleavage surfaces in the tetragonal system. Symmetry requires that the {hhl} and {h0l} cleavages yield four differently oriented surfaces. The {hkl} cleavage yields eight differently oriented surfaces. (B) Hexagonal cleavages that encounter both the c and a axes. Depending upon the particular cleavage and crystal class, three, six, or twelve differently oriented surfaces may be produced. (C) A grain with pyramidal, rhombohedral, or scalenohedral cleavage, as viewed in vertical section. All of these cleavage types cause the c axis to be oriented at an oblique angle to the stage. In plan view, the grain outline is that of a parallelogram. At extinction, adjacent cleavage edges are symmetric to the NS and EW cross hairs.

basal cleavages yield only values of ω. Rhombohedral, pyramidal, or scalenohedral cleavages yield values of ε' and ω. A prismatic cleavage is more useful than any other when working with fragments, as the grains furnish a measurement of the true value of ε (rather than ε') as well as ω; furthermore (assuming the lack of a basal cleavage), the directions along which ε and ω are measured relative to the grain shape are also known, as the grains are elongate parallel to the c axis (Fig. 5-4B). It is a simple matter to rotate the grain into an extinction position, putting the c axis parallel to the vibration direction of the lower polarizer, and make a Becke test for ε. As the ε and ω vibration directions within the grains are known, the optic sign is quite easily determined as well by comparing the two indices.

Some minerals may possess more than one type of cleavage. For example, a particular mineral may have both a good {001} and a perfect {110} cleavage. In this mineral, most of the cleavage fragments lie on {110} surfaces and present a roughly rectangular outline with some elongation parallel to {110} (Fig. 5-5). Extinction is parallel to both the prismatic and basal edges. A few grains lie on the {001} cleavage and have a rectangular or square outline (due to the {110} cleavages) and remain in extinction during stage rotation. Any combination of cleavages is possible, including two of the same type, such as {111} and {112}.

INTERFERENCE EFFECTS WITH MONOCHROMATIC LIGHT

The origin of interference effects is best understood by first considering the effects of monochromatic light on a uniaxial crystal. Figure 5-6 shows a grain on the microscope stage, oriented so that its two privileged vibrational planes are diagonal to the cross hairs. As the plane-polarized light from the lower polarizer enters the grain, it is constrained to vibrate within the two privileged vibration planes. As the wavelength, index of refraction, and velocity are different for the light

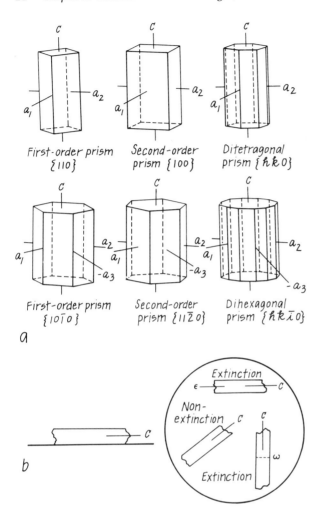

Figure 5-4
(A) Prismatic cleavage directions in tetragonal and hexagonal materials; the upper and lower basal pinacoid surfaces are shown only to emphasize the orientation of the crystals. Note that all of the prismatic cleavages are parallel to the c axis. Two, three, four, or six differently oriented but symmetrically equivalent surfaces are produced. (B) Vertical section of a grain lying on a prismatic cleavage; at least one additional cleavage surface is parallel to the c axis. In plan view, the grains are elongate. Extinction occurs when the cleavage edges are parallel to the NS or EW cross hairs. When extinction occurs with EW elongation, the ε index can be measured; when extinction occurs with NS elongation, the ω index can be measured.

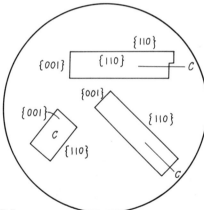

Figure 5-5
A material with a perfect {110} cleavage and a good {001} cleavage. The grains lie mainly on {110} surfaces, have parallel extinction, and are elongate parallel to the *c* axis. The smaller cleavage edge, {001}, is perpendicular to the *c* axis. A small number of grains lie on the {001} cleavage; these grains tend to be blocky in outline and exhibit extinction in all positions of rotation of the stage.

within each of these planes, the slow wave lags behind the fast ray. The retardation of one wave behind the other results in a variety of different types of polarized light when the two waves are combined upon leaving the crystal.

Full-Wave Retardation

In Figure 5-7A, a mineral grain on the stage has been eliminated and in its place are two diagonal lines, which represent the planes of the privileged directions within the grain. The incident plane-polarized light, in accordance with the world standard, vibrates within the EW plane. This wave constantly oscillates left and right as it moves up to the mineral grain. When this wave is extended (with maximum amplitude) to the right we will regard this as a crest and when extended to the left, a trough. Figure 5-7B shows what happens as the plane-polarized wave enters the crystal. The single wave is resolved into two perpendicular components, done graphically on the

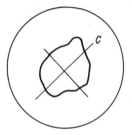

Figure 5-6
A grain with its privileged vibration directions oriented diagonally.

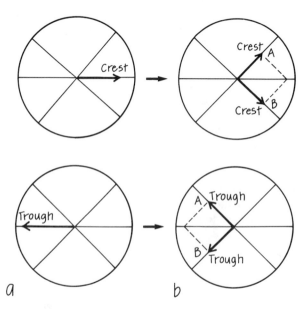

Figure 5-7
(A) The privileged vibration directions of a grain are diagonally oriented. Plane-polarized (EW) light arrives at the grain. (B) The incident light is resolved into the two privileged vibration directions within the grain by constructing lines (dashed) from the incident wave position normal to both privileged directions.

figure by drawing dashed lines from the EW polarized vibrational plane, perpendicular to the two diagonal planes. The intersection points (A and B in Fig. 5-7B) indicate the initial amplitudes of the two waves that develop within the two component planes. For descriptive purposes the NE

Figure 5-8
(A) Incident plane-polarized EW light (OP) enters a grain and is resolved into the two diagonal privileged directions, OA and OB. (B) The light that vibrates in the NW–SE plane arrives at the top of the grain with the same amplitude and vector direction as when it entered after two complete oscillations. (C) The light that vibrates in the NE–SW plane arrives at the top of the grain with the same amplitude and vector direction as when it entered after a single oscillation. (D) The two vibrations (EC and ED) at the top of the grain are in identical relationship to their entering positions (OA and OB), thus the resultant vibration EF is identical in orientation and amplitude to the original vibration OP. A retardation of an integral number of wavelengths (1λ, 2λ, 3λ, etc.) restores the incident plane-polarized light to its initial plane of polarization.

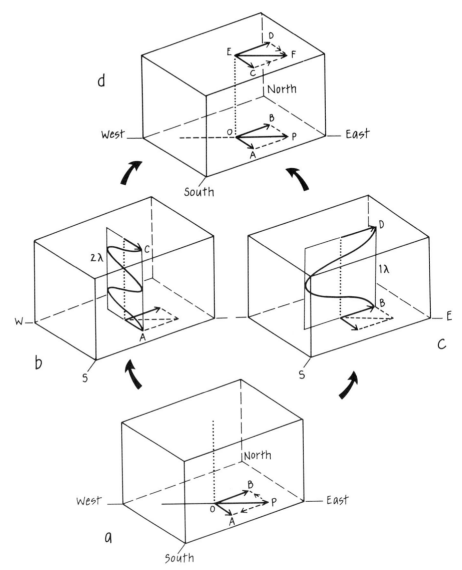

and SE vibration directions are called crests (and the NW and SW, troughs).

What might happen within the crystal is shown in Figure 5-8. In part A, at the base of the crystal block, incident plane-polarized light, shown as line segment OP, enters the crystal and is resolved into the two privileged vibrational planes OA and OB. Part B shows that light within the vertical plane that contains OA assumes a particular wavelength and moves to the top of the crystal. For this ray, the crystal is exactly 2 wavelengths thick; upon reaching the top of the crystal, this light wave has the same amplitude and direction of vibration as when it entered the crystal, and point C is directly over point A. This situation occurs when the crystal thickness is equivalent to any integral multiple of the wavelength (2λ, 3λ, 4λ, etc.). The light within the vertical plane that con-

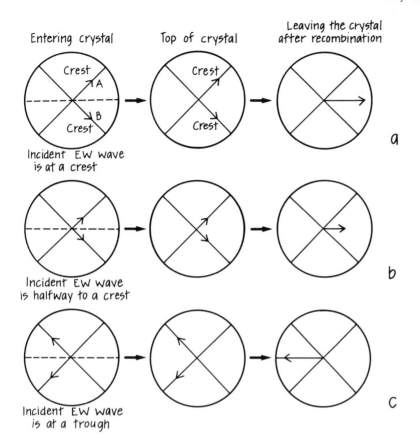

Figure 5-9
A schematic view of full-wave retardation (1λ, 2λ, 3λ, etc.). The vibration directions within the grain are NE–SW and NW–SE, and the incident plane-polarized light is in the EW plane as usual. (A) The incident light enters in a crest position (vector direction to the right), is resolved into the two privileged directions at A and B, and recombines at the top of the grain into an EW plane-polarized vibration. (B) The entering light, halfway between a crest and null position, recombines into an EW plane-polarized vibration. (C) The entering light, in a trough position, recombines into an EW plane-polarized vibration.

tains OB is shown in part C; for this ray, the grain thickness is equivalent to a single wavelength. This wave also arrives at the top of the crystal with the same amplitude and direction of vibration as when it entered, and point D is directly over point B.

The block in part D shows the two vibrations at the top of the crystal. As the two waves leave the crystal, they recombine into a single vertical plane of vibration EF because the crystallographic vibrational constraints have been eliminated.[1] A method of recombination, shown graphically in part D, is to construct a parallelogram with the two wave amplitudes EC and ED as adjacent sides; the single resultant plane of vibration produced is diagonal EF of the parallelogram. This plane of vibration EF is parallel to the plane OP of the incident light below the crystal, which indicates that the original direction of polarization was unchanged by the crystal in spite of the fact that the slower vibration was retarded by a whole wavelength while going through the crystal.

We obtain the same result if the incident light enters the crystal at a different stage in its vibration. Figure 5-9 shows three examples of full-wave retardation: the first is that of Figure 5-8, the second shows an incident wave at half-crest amplitude, and the third shows an incident wave at the trough position. In each the incident light

---

1. Many texts on optical mineralogy state that the two vibrations produced by an anisotropic crystal are not combined until they reach the upper polarizer. Although incorrect, such a description is easily visualized, and it leads to correct conclusions about the resultant vibration directions after the light has left the upper polarizer.

Figure 5-10
A schematic view (similar to Fig. 5-9) of the effect of grain retardation that is an odd multiple of a half wavelength ($\frac{1}{2}\lambda$, $\frac{3}{2}\lambda$, $\frac{5}{2}\lambda$, etc.). Combination of vibrations upon leaving the grain results in plane-polarized NS light. The crystal has rotated the incident plane-polarized light by 90°.

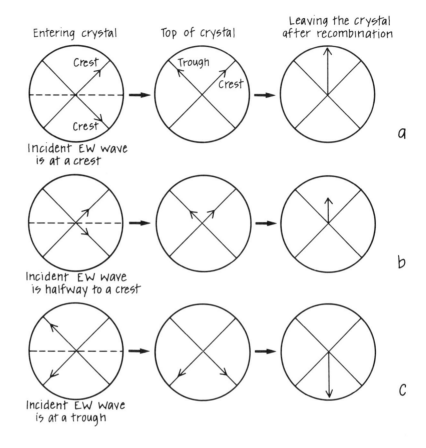

emerges in the same vibration plane it had as it entered. Other similar cases can be constructed, and would yield similar results. Thus we can conclude that if the thickness of the crystal (and the velocity difference of the two vibrations) is such as to produce a retardation of whole numbers of wavelengths, plane-polarized light emerges from the top of the crystal in its original plane of polarization.

Half-Wave Retardation

What if the crystal in Figure 5-8 were half as thick? In reducing the thickness by half, the retardation between the two vibrations would be reduced by half, to $\frac{1}{2}\lambda$. We will discuss here all odd multiples of half-wave retardation ($\frac{1}{2}\lambda$, $\frac{3}{2}\lambda$, $\frac{5}{2}\lambda$, etc.).[2]

The results of an odd multiple of $\frac{1}{2}\lambda$ retardation are shown in Figure 5-10. In part A, the incident plane-polarized light (at a crest) is resolved into two planes of vibration. When the two waves reach the top of the crystal, they are not in similar crest positions. By being retarded $\frac{1}{2}\lambda$, one of these waves is in a trough position. When the two waves combine at the top of the crystal, the single

---

2. It should be noted that some authors refer to fractional separation of waves as *phase difference*, reserving the term *retardation* strictly to the total linear separation of the two waves. Thus a separation of two waves by $2\frac{1}{2}\lambda$ would have a phase difference of $\frac{1}{2}\lambda$ and a retardation of $2\frac{1}{2}\lambda$.

wave produced is located in the NS plane. Examination of a few other situations (Fig. 5-10B and C) reveals that in spite of differences in the amplitude and vector direction of the incident EW plane-polarized light, the single wave produced at the top of the crystal remains in the NS plane. Thus, a crystal of a thickness that produces a half-wave retardation rotates transmitted light 90° from its original plane of vibration; EW plane-polarized light is converted to NS plane-polarized light.

Quarter-Wave and Intermediate Retardations

Consider what happens if a crystal causes a $\frac{1}{4}\lambda$ retardation (Fig. 5-11): the incident light becomes two waves that move to the top of the crystal, but when one of the two waves has reached a crest, the other is only halfway between a crest and a trough. Figure 5-11 shows that recombination produces a single vibration plane that changes in direction; note that although the direction of polarization changes, the amplitude of the resultant wave remains the same. As described in Chapter 2 and shown in Figure 2-11, the type of light produced is called *circularly polarized light*.

The most common situation of all occurs when a crystalline material produces a retardation that is not $\frac{1}{4}\lambda$, $\frac{1}{2}\lambda$, 1λ, or any equivalent multiple. Such retardations will yield a resultant vibration that changes regularly both in orientation and in amplitude. The amplitudinal change follows an elliptical pattern. This is called *elliptically polarized light*.

Figure 5-12 summarizes the above discussion using a diagonally oriented crystalline wedge shown in vertical section on the microscope stage. Increasing thickness is related to increase in retardation, which in turn controls the type and direction of the polarized light produced.

The effect of plane-polarized monochromatic light on an anisotropic grain viewed between crossed nicols is shown in Figure 5-13C. The grain has an irregular lensoid shape and shows alternate bands of light and darkness. As the

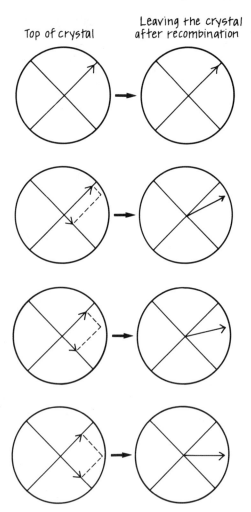

Figure 5-11
A schematic view (similar to Fig. 5-9) of the effect of grain retardation that is an odd multiple of a quarter wavelength ($\frac{1}{4}\lambda$, $\frac{3}{4}\lambda$, $\frac{5}{4}\lambda$, etc.). The two vibrations arrive at the top of the crystal in a constantly changing pattern of relative amplitudes and vector directions. The resultant vibration constantly changes in direction while maintaining a constant amplitude, a vibrational pattern called circularly polarized light. If the retardation is not a multiple of $\frac{1}{4}\lambda$, $\frac{1}{2}\lambda$, or 1λ, a vibration pattern would develop that constantly changes in direction while changing amplitude in an elliptical pattern; this is called elliptically polarized light. Note that in the upper left diagram, the NW–SE wave is at a null position and is therefore not shown.

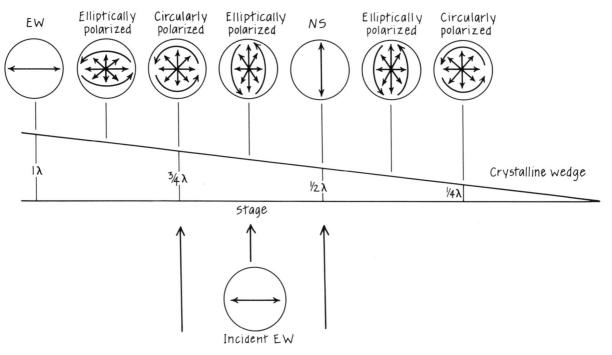

Figure 5-12
Vertical section of an optically anisotropic wedge, diagonally oriented on the microscope stage. The nicols are crossed and monochromatic light is assumed. The type of polarized light produced for different wedge thicknesses is indicated above the wedge.

edges of the grain are typically thinner than the center, it is possible to contour the grain with lines indicating a variety of grain thicknesses. These contours are also lines of constant retardation within the grain (Fig. 5-13A). The grain has been contoured for thicknesses equivalent to $\frac{1}{2}\lambda$ retardation, 1λ retardation, and $\frac{3}{2}\lambda$ retardation. For each of these contours, the type of polarized light that emerges is indicated in Figure 5-13B. Areas between the indicated retardation levels produce circularly or elliptically polarized light. Light from this grain that reaches the upper polarizer is either transmitted or eliminated as a function of the types of polarization produced by the different amounts of retardation.

The upper polarizer transmits light (or a component of it) that is vibrating in the NS plane. Circularly and elliptically polarized light have components in all directions, so at least part of this light passes through the upper polarizer. Light that has been retarded by half-waves vibrates in the NS plane and is also able to pass through the upper polarizer. Light that has been retarded by full waves is eliminated by the upper polarizer and appears as black bands within the crystal. The color of the light that emerges from the grain is that of the monochromatic source—for example, if a sodium vapor lamp is used, the grain exhibits alternating black and orange bands (Fig. 5-13C).

INTERFERENCE EFFECTS WITH POLYCHROMATIC LIGHT

When viewed in polychromatic light between crossed nicols, anisotropic crystals show colors in most orientations. These are the same colors that are observed in oil slicks on a wet street, or the iridescent colors of a starling's feathers, or within mother-of-pearl. All result from the interference of light waves. Most microscopic examination uses a polychromatic light source. As a polychromatic light source is merely a collection of monochromatic sources, we can first consider whether each type of monochromatic source operates in an identical manner.

First assume a material that has uniform dispersion for all wavelengths of light—that is, the difference between ε and ω is identical for all wavelengths. This assures us that the retardation is the same for each wavelength. The retardation (or path difference) within a crystal can be determined by use of the equation

$$\Delta = d(n_2 - n_1)$$

where Δ is the retardation, d is crystal thickness, n_2 is the larger index of refraction, and n_1 is the smaller one. A large index difference means a large velocity difference (and vice versa); a large velocity difference favors greater relative retardation. Similarly, a thicker grain produces more retardation than a thinner grain of the same material. As an example, assume that we have a grain 0.02 mm thick; in terms of wave notation this is 0.02×10^6 nm. If the indices of refraction for the wavelength used are 1.587 and 1.572, the equation permits determination of path difference, that is, the retardation:

$$\Delta = d(n_2 - n_1)$$
$$= 0.02 \times 10^6 (1.587 - 1.572) \text{ nm}$$
$$= 300 \text{ nm}$$

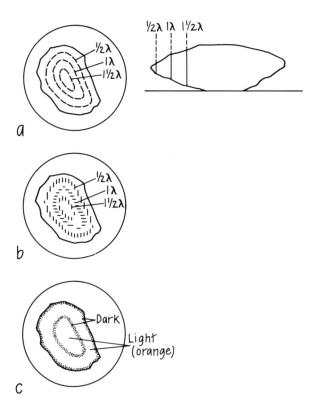

Figure 5-13
An anisotropic lenticular grain is diagonally oriented on the microscopic stage. Monochromatic orange light is assumed. (A) The grain is contoured for thicknesses that produce particular amounts of retardation. (B) Those parts of the grain that produce a half-wave retardation yield NS plane-polarized light. Retardation of an integral multiple of 1λ yields EW plane-polarized light. Intermediate retardations yield circularly or elliptically polarized light. (C) The upper polarizer permits transmission of plane-polarized NS light (or components of this direction); the microscopist sees light of an orange color. Light vibrating in the EW vertical plane (or components in this direction) is eliminated by the upper polarizer. The very thin edge of the grain produces almost no retardation and appears almost black.

Figure 5-14
Vertical section of an optically anisotropic wedge in diagonal orientation on the microscope stage. After transmission through the wedge, monochromatic light reaches the upper polarizer. (A) The intensity of incident red light (λ = 700 nm) within the wedge is at a maximum when wedge thickness is $(n/2)(700\ nm)$, where n is an odd number. (B) The intensity of incident violet light (λ = 400 nm) within the wedge is at a maximum when wedge thickness is $(n/2)(400\ nm)$, where n is an odd number. When incident light is dichromatic (red plus violet), the colors developed are a combination of both the red and violet intensity curves. The only region of darkness is at the thin edge of the wedge.

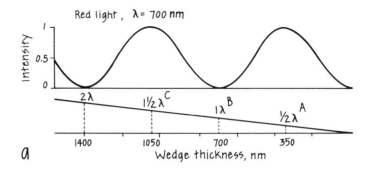

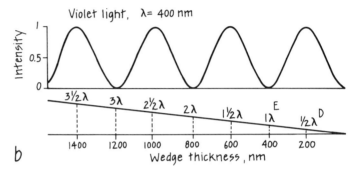

To describe the retardation in terms of multiples or fractions of wavelengths, merely divide the retardation by the wavelength. If the substance is illuminated by light having a wavelength of 600 nm, the calculated retardation is one-half wavelength.

We shall first determine the effect of using two different wavelengths of light. A single-crystal anisotropic wedge is placed on the stage of the microscope in the diagonal position between crossed nicols. Red light (λ = 700 nm) is transmitted through the wedge (Fig. 5-14). At point A on the wedge, the thickness is sufficient to produce a retardation of 350 nm, or $\frac{1}{2}\lambda$ retardation for red. At point B, the retardation is 700 nm, and the retardation is 1λ, and so on. Looking at the crystal wedge through the microscope with the nicols crossed, we would see darkness at the thin edge, changing to maximum intensity of red at point A, a gradual decrease to black at B, and a rise to a maximum red intensity at C.

Now replace the red light with a violet light (λ = 400 nm). When the crystal wedge thickness is sufficient to produce a 200 nm retardation (at D), violet light has a retardation of $\frac{1}{2}\lambda$. When the wedge thickness is equivalent to 400 nm retardation (at E), the path difference is 1λ, and so on.

Figure 5-14 shows how the red and violet light intensities vary as a function of wedge thickness. Note that the high and low intensity positions of the red and violet light do not overlap; a low intensity of one color may correspond to a high intensity of the other color. Hence if both red and violet light are transmitted through the wedge simultaneously, the black bands are largely eliminated, except at the thin edge where retardation is minimal for all wavelengths. We see a series of bands of various shades and intensities of red and violet.

A composite diagram for all of the interference colors of a normal polychromatic light source can be constructed. The result is total elimination of all dark bands (except at the very thin edge of the wedge) and a sequence of all of the colors of the spectrum. For grain thicknesses where one or more colors are eliminated, the remaining colors (in various intensities) are transmitted. A variety of partially depleted spectrums is observed.

The sequence of interference colors, shown in Plate 1 and known as the *Michel-Lévy Chart* or *interference color chart*, is somewhat repetitious; consequently, it is arbitrarily divided into subdivisions called *orders*; first-order colors extend from black through the first red, second-order colors from the first red through the second red, and so forth. Each new order begins with an additional 550 nm of retardation. These subdivisions were chosen because the adjacent colors at 550-nm intervals are extremely sensitive to small changes in retardation. Observe that blue occurs only in the second and third orders. Above the third order, green and red colors alternate. As the sequences rise to higher orders, there is a tendency for greater amounts of overlap of individual colors, so that the pastel colors of the higher orders become milky white. This white is called *white of the higher order*, in order to contrast it with the white found within the first-order colors.

ACCESSORY PLATES AND WEDGES

Optical mineralogy relies on the use of one or more accessory plates. All accessories are inserted into the microscope through the accessory slot just above the objective lens, and when inserted, the privileged directions of the accessory plate are in the diagonal position (NE–SW and NW–SE). When used with grains viewed between crossed nicols, insertion of an accessory plate causes a change in the level of interference colors.

The purpose of accessory plates is to furnish information on the nature of the vibration directions within anisotropic grains or interference figures, and in so doing, to facilitate index of refraction and optic sign determinations.

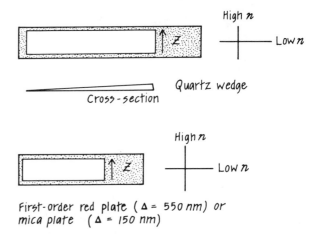

Figure 5-15
Commonly used accessory plates. The arrow is conventionally parallel to the short edge of the accessory, and indicates the vibration direction having the higher index of refraction (and slower velocity).

The *quartz wedge* is a wedge-shaped slice of quartz that is glued between two pieces of glass, and held within a metal or plastic holder. The quartz is cut from a single crystal. The privileged directions of the quartz are oriented parallel and perpendicular to the long direction of the holder. The direction of the higher (slower) index is usually indicated with an arrow (and the letter Z or γ), and in *most* wedges is parallel to the short direction of the wedge (Fig. 5-15). When the thin edge of the quartz wedge is continuously inserted between crossed nicols, it produces increasingly higher retardation as successively thicker amounts of quartz enter the light path. This yields the sequence of interference colors seen in Plate 1. As the accessories are not located in the focal plane of the objective lens, the observer views only the interference colors and not the accessory itself.

The *first-order red plate* (often called a gypsum plate) consists generally of a single crystal of gypsum (or less commonly quartz), ground to constant thickness, glued between glass, and put in a metal or plastic holder. The privileged direc-

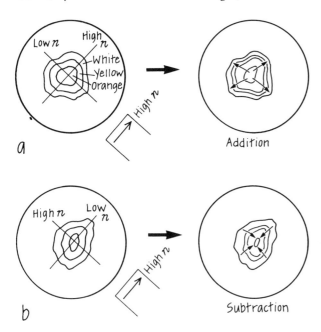

Figure 5-16
A diagonally oriented grain viewed between crossed nicols. The quartz wedge enters from the SE. (A) When like vibration directions are parallel in the grain and the quartz wedge, retardation is additive. This is seen on the grain as an outward flow of interference colors during insertion of the quartz wedge. (B) When unlike vibration directions are parallel in the grain and the quartz wedge, retardation is subtractive. This is seen on the grain as an inward flow of interference colors during insertion of the quartz wedge.

tions of the crystal are oriented similar to the quartz in the quartz wedge. An arrow parallel to the short direction of the holder indicates the high index (slow) direction (Fig. 5-15). The thickness of the plate is such as to furnish a retardation of 550 nm. When inserted between crossed nicols with no specimen on the stage, a first-order red interference color covers the field of view.

The *mica plate* is a single crystal of muscovite, of a thickness that produces a retardation of 147 nm (gray color); as this is about one-fourth the wavelength of sodium light, this accessory is often called a quarter-wave plate.

USING ACCESSORIES TO DETERMINE VIBRATION DIRECTIONS IN IMMERSION MOUNTS

Assume that you wish to measure the indices of refraction of an anisotropic mineral fragment in an immersion mount. The first part of the procedure is to find the vibration directions that correspond to the high and low indices. Up to this point it has been necessary to rotate the grain into each of its extinction positions, uncross the nicols, and then compare the results of Becke tests. This is relatively simple with immersion liquids but may be very difficult in thin section. An accessory plate solves the problem easily.

The procedure is straightforward. Cross the nicols. Rotate the stage until the grain is about 45° from its extinction position; that position is approximated by rotating the stage until the interference colors of the grain are at their brightest. If the grain shows several interference colors, insert the quartz wedge; alternatively, if the grain shows only the first-order white or gray it is preferable to use either the first-order red or mica plate. In this example we will use the quartz wedge (Fig. 5-16).

The vibration directions of the accessory are in the diagonal position and should be parallel to the vibration directions of the grain. Insertion of the accessory changes the interference colors within the grain; the resultant interference colors depend upon a combination of the retardation produced by the grain itself and the retardation produced by the accessory. This combination may be one of either addition or subtraction.

Figure 5-16A shows a situation in which the slow (high) index direction of the quartz wedge is parallel to the slow (high) index direction of the crystal. The fast (low) index directions are, of course, also parallel. The retardation within the grain is reinforced by the retardation within the accessory. In effect, the fast vibration direction of the grain gets a second chance to become fast in the quartz; the separation (retardation) between the two vibrations is increased as compared to the situation when the accessory was not present. An increase in retardation means an increase in interference colors (toward high orders). Thus higher level color bands appear to flow from the thicker center of the grain toward the lower level colors of thinner edges as the quartz wedge is continuously inserted (thin to thick end). Examination of any part of the grain reveals a continuous rise in interference colors. This situation is one of *addition*.

In the opposite situation (Fig. 5-16B), where the fast and slow index directions of the quartz are opposed to their counterparts in the grain, the effect of the accessory is to decrease the overall retardation by *subtraction* or compensation. The interference color at any point on the grain decreases toward lower orders during insertion of the accessory. This is seen as a flow of color bands from the thin edge of the grain toward the thicker center.[3]

Use of a quartz wedge, therefore, is a simple and rapid method for determining the nature of the vibration directions within the grain. The next thing to do is a Becke test measurement of a particular index direction. The grain is rotated to align the desired index direction with the privileged direction of the lower polarizer. In this case (Fig. 5-16A) the high index direction is to be measured, so the grain is rotated 45° clockwise to extinction; the upper polarizer is removed and the Becke test performed. Alternatively, the low index direction of the grain is brought to the EW measuring position by a 45° counterclockwise rotation to extinction. When unlike directions in the grain and accessory are parallel, the procedure is reversed.

If the mineral grain under examination shows few or no bands of interference colors and instead appears as some shade of white or gray, it is generally easier to use the first-order red or mica plate for determination of high and low index directions. Assume that the grain has mainly a first-order white interference color. As in the previous case, the stage is rotated until the grain is in the 45° position, placing its privileged directions parallel to those of the accessory.

Assume first that we are using the first-order red plate, and that like vibrations are parallel in the grain and the plate. The first-order red plate produces a retardation of 550 nm; the white part of the grain has a retardation of about 158 nm (see Plate 1). With like directions parallel in both the grain and the plate, the retardation of the two are added together, producing an overall retardation of 708 nm. Referring to the retardation scale on the left side of Plate 1, it can be seen that the interference color produced is a bluish green—a higher order color than that of either the first-order red plate or the grain. Such a rise in color indicates that like directions correspond in both grain and plate. When unlike directions are parallel, the effect is one of compensation or subtraction (550 − 158 = 392 nm), and the resultant grain interference color is yellow. With use of the mica plate, addition of 147 nm (retardation of the mica plate) and 158 nm (retardation of the grain) yields an overall retardation of 305 nm, which would result in an orange-white color. Alternatively, if unlike vibrations were parallel, the two retardations would be subtractive, producing an overall retardation of 11 nm; the grain would appear to be essentially black.

If there is any doubt as to whether the accessory test is additive or subtractive (due to indeterminate color effects), the grain can be rotated 90° (into an opposite diagonal position) after insertion of the accessory. Color rise or fall during rotation of the stage will reveal whether the original position was additive or subtractive. It may be useful, when trying this procedure for the first few times, to match the color changes with those on the interference color chart; after becoming familiar with the color sequence this will not be necessary.

---

3. It is possible to overcompensate the interference colors of a grain. Insertion of a quartz wedge can not only reduce the interference colors to black, but continue past that point, such that the colors rise to higher levels with continued insertion. The direction of color band movement, however, remains the same (from the thinner edge toward the thicker center).

USING ACCESSORIES TO DETERMINE VIBRATION DIRECTIONS IN THIN SECTION

The procedure for determining the high or low index directions of a grain in thin sections is identical to that used for a crushed fragment in an immersion liquid; the only difficulty lies in the fact that many grains in thin sections are of uniform thickness and do not possess gradually sloping edges. The result is that most grains possess a uniform interference color throughout, so the typical color bands seen in crushed grains are absent. When the accessory is inserted, it is sometimes difficult to determine whether the interference colors have risen or fallen. It is sometimes also difficult to determine the order of the color originally present in the mineral.

The difficulty is usually resolved easily by using a quartz wedge. After rotating the grain into the 45° position between crossed nicols, the quartz wedge is slowly inserted and the sequence of colors is noted. By reference to Plate 1, it is possible to determine whether the sequence is rising toward higher orders or decreasing toward lower orders. If it is desired to know the particular order of a color exhibited by a mineral grain, the grain can be rotated into a subtractive position and the sequence of colors noted during insertion of the wedge; the colors will decrease toward zero retardation (black). Taking note of the sequence of colors to reach complete compensation will reveal the original order of color. If, however, the interference colors of the grain are higher than the fourth order, complete compensation cannot be achieved with the quartz wedge.

PLEOCHROISM

As with isotropic materials, uniaxial crystals may exhibit selective absorption of light. This results in the crystal being colored. The color of a uniaxial crystal may be the same for all vibration directions, or it may vary according to the vibration direction.

Consistency of absorption can be determined by observing a grain with uncrossed nicols during stage rotation; the grain must be oriented such that the c axis is not normal to the stage. If the grain exhibits *differential absorption* or *pleochroism*, its color will change during rotation of the stage—either in intensity, or from one color to another.

If the mineral displays differential absorption, it is possible to relate each color to a particular vibration direction. Ideally a grain should be obtained in which the c axis is parallel to the stage; this permits the particular colors to be related to the ε (rather than ε') and ω vibration directions.

Assume a tetragonal material with perfect {110} prismatic cleavage. Fragments are elongate in the direction of the c axis. With crossed nicols, rotate the stage until the elongation direction is parallel to the privileged vibration of the lower polarizer; at this point the grain becomes extinct, and the single vibration direction within the fragment yields an ε index of refraction. Uncross the nicols and note the color of the grain. Then cross the nicols and rotate the stage 90° to the next extinction position. The direction of elongation is now NS, and the single vibration direction within the fragment yields an ω index of refraction. Uncross the nicols and note the color.

If the grain changed in intensity of color, this could be written as an *absorption formula*. For example, if the ω vibration direction yielded a dark green color and the ε vibration direction a light green color, the absorption formula could be written as "Abs. $\omega > \varepsilon$" or "Abs. $O > E$" (where O refers to the vibration direction of the ordinary index ω and E is the vibration direction of the extraordinary index ε). If the mineral changed color during stage rotation (Fig. 5-17), this could be written as a pleochroic formula, such as "ω = green, ε = yellow" or "O = green, E = yellow." Knowledge of the pleochroic formula or absorption formula is often useful in distinguishing among various similar unknown materials.

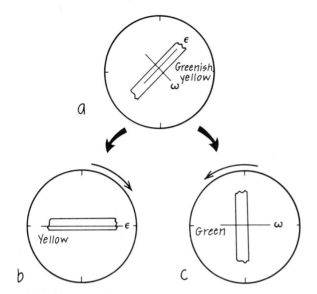

Figure 5-17
The determination of a pleochroic formula. (A) A colored crystal fragment with perfect {110} cleavage is found. The elongation direction is parallel to c. (B) The fragment is rotated into extinction, with its elongation parallel to the privileged direction of the lower polarizer. The color associated with the ε vibration direction is noted after uncrossing the nicols. (C) The fragment is rotated 90° from position (B). With uncrossed nicols, the color associated with the ω vibration direction is noted.

ADDITIONAL READINGS

Battey, M. H. 1972. *Mineralogy for Students.* London: Longman, 29–139.

Berry, L. G., and B. Mason. 1959. *Mineralogy.* San Francisco: W. H. Freeman, 185–209.

Bloss, F. D. 1961. *An Introduction to the Methods of Optical Crystallography.* New York: Holt, Rinehart and Winston, 92–150.

Ford, W. E. 1922. *A Textbook of Mineralogy* (E.S. Dana). New York: John Wiley & Sons, 720 pp.

Winchell, A. N. 1947. *Elements of Optical Mineralogy: Part I, Principles and Methods,* 5th ed. New York: John Wiley & Sons, 1–36, 104–138.

6

Uniaxial Interference Figures

He was a bold man that first ate an oyster.
JONATHAN SWIFT

The arrangement of light rays used to measure indices of refraction, observe cleavages, and so forth, is referred to as *orthoscopic illumination;* the light rays move in essentially vertical parallel paths through the lens system of the microscope and permit observation of the grain image. To view interference figures, however, the light must be converted to *conoscopic illumination.* Here the light rays converge within the specimen and diverge upon leaving the point of convergence. Light convergence is created by using an auxiliary condensing lens that is either rotated or raised into operating position above the substage. Instructions furnished with the instrument will describe which procedure is used with your particular microscope.

In order to bring a significant amount of the divergent light rays into the microscope tube, it is necessary to have the base of the objective lens as close as possible to the mineral grain. This requires use of a high-power objective (40× or greater) of high numerical aperture, which has its focus just above the cover glass of the specimen slide.

An image of light and dark patterns, called an *interference figure,* is formed on the upper curved surface of the objective lens. This image can be viewed either by removing the ocular (and perhaps substituting a pinhole eyepiece if available), or by leaving the ocular in place and inserting the Bertrand lens. The pinhole eyepiece is simply a metal disk with a small hole in the center. Its purpose is to prevent stray light from entering the microscope tube and to keep the observer's line of sight in the center of the tube. The Bertrand lens brings the upper surface of the objective lens into focus within the ocular. A large and perhaps somewhat diffuse interference figure will be seen with the Bertrand lens. Both alternatives should be tried with your microscope in order to determine which produces the better image.

The steps required to produce an interference figure are as follows:

1. With the nicols crossed, find an appropriate grain with the low- or medium-power lens. An ideal grain, in immersion mounts, is one that is at least slightly isolated from its neighbors, is of average size, and possesses the desired level of interference colors; the level of interference colors relates to the orientation of the grain, which determines the type of interference figure obtained (as explained below). When working with thin sections, the better interference figures are obtained with grains that are sand size (0.06 mm) or larger. Smaller grains, however, should be checked as well, in order to determine the limiting minimum size that can be used with your

Plate 2
Interference figures and sign reactions.

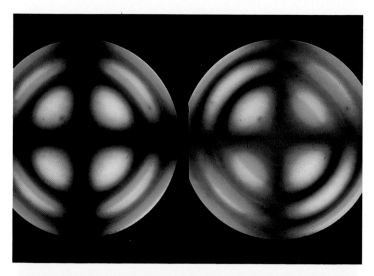

1. (Left) A uniaxial optic axis interference figure. (Right) With insertion of a first-order red plate, the optic sign is established as positive (+).

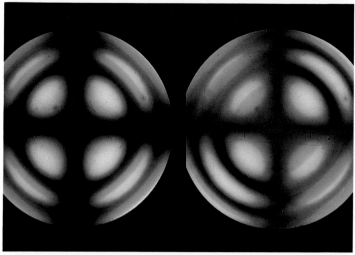

2. (Left) A uniaxial optic axis interference figure. (Right) With insertion of a first-order red plate, the optic sign is established as negative (−).

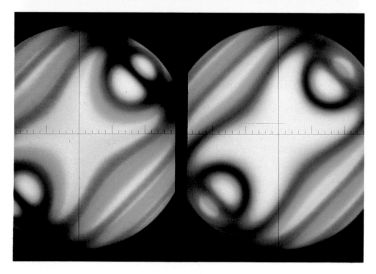

3. (Left) A biaxial acute bisectrix interference figure in the diagonal position. (Right) With addition of a first-order red plate, the colors have decreased between the melatopes, indicating that the optic sign is positive (+).

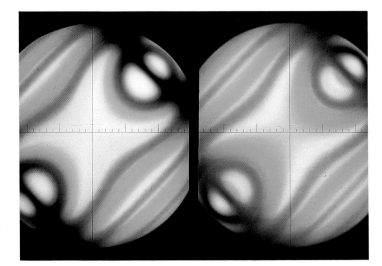

4. (Left) A biaxial acute bisectrix interference figure in the diagonal position. (Right) With addition of a first-order red plate, the colors have increased between the melatopes, indicating that the optic sign is negative (−).

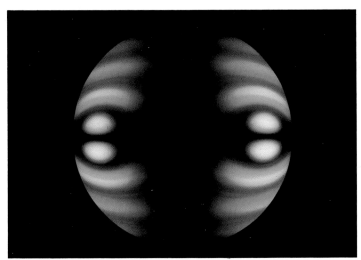

5. A biaxial acute bisectrix interference figure in the cross position. The melatopes are located along the EW isogyre. Notice that the NS isogyre is much broader than the EW isogyre. (Figures 1 through 5 from J. Hinsch, E. Leitz, Inc.)

microscope. Again, the level of interference colors should be considered if a particular type of interference figure is desired. In general, the most information is obtained on anisotropic grains with relatively low interference colors, as the optic axes of these grains are near-parallel or parallel to the microscope's lens system.

2. After locating an appropriate grain, put the high-power objective into place and insert the upper polarizer. Be sure that the grain is properly focused and remains centered on the intersection of the cross hairs during stage rotation.

3. Open the aperture diaphragm as much as possible.

4. Insert the condensing lens. Be sure that it is raised to its maximum height, almost touching the base of the slide; if too low, convergence will occur below the specimen.

5. View the interference figure either by inserting the Bertrand lens or by removing the ocular and looking directly down the open microscope tube (or looking through a pinhole eyepiece).

Any anisotropic grain in any orientation will produce some kind of interference figure. We shall consider in turn three types of uniaxial interference figures that are produced when the c axis is perpendicular, inclined, or parallel to the microscope stage.

UNIAXIAL OPTIC AXIS FIGURES

The c axis of a uniaxial crystal corresponds to its optic axis. When the c axis of a crystal is perpendicular to the stage, the viewer is looking along the optic axis. Mineral grains in this orientation are easily recognized as they remain in extinction when rotated between crossed nicols. The interference figure obtained for a crystal in this orientation is called an *optic axis figure* (see left part of Figs. 1 and 2 in Plate 2).

The origin of the uniaxial optic axis figure can be explained with the aid of Figure 6-1. Part A shows a uniaxial crystal in the field of view whose c axis is perpendicular to the stage. Consider cross sections within this crystal along vertical planes that are oriented EW, NS, and NE–SW. For the sake of simple explanation, imagine that the interference figure originates at a single point of intersecting convergent light within the grain (at a point such as X within the cross-section of parts B, C, and D); in fact every point within the grain represents such a convergent light intersection.

Any randomly chosen vertical plane through the crystal in Figure 6-1A is parallel to the crystal's c axis, and is therefore a principal plane. (Recall that a uniaxial crystal permits vibration within a principal plane and in a plane perpendicular to it.) Therefore, permitted vibrations are parallel and perpendicular to any randomly chosen vertical plane. Which of these vibrations may be utilized by the crystal is a function of the direction of the plane-polarized light from below.

Consider first what happens relative to the EW cross-section (part B). The incident light from the substage is plane-polarized in the EW plane, and the condensing lens converges this light to the point X within the grain. The condensing lens makes the ray paths converge to a point but does not change the type of polarization or the orientation of the plane of polarization; after convergence, this light continues to be plane-polarized in the EW vertical plane. This plane, coincident with the plane of the page, is also a principal plane within the crystal. As the principal plane is one of the privileged vibration planes, the EW plane-polarized light travels through the crystal without becoming doubly polarized and leaves the crystal plane-polarized EW. It is transmitted through the crystal with an index of refraction that has some value of ε'; the only exception is the single ray that is parallel to the crystal's c axis, which has the index ω.

Consider now what occurs within a NS vertical section (part C). The vibration direction of the incident EW plane-polarized light is perpendicular to the NS section. As the NS plane, too, is a prin-

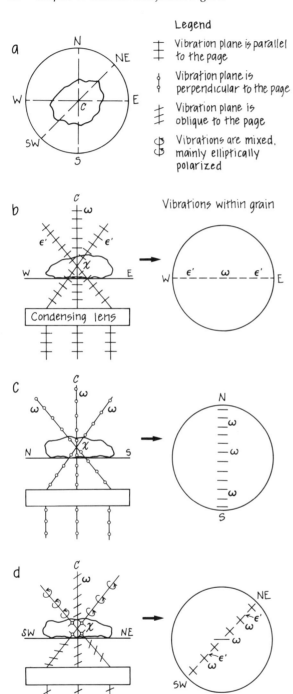

cipal plane, it follows that the EW vibration direction of the incident light, being perpendicular to this principal plane, corresponds to a privileged vibration direction within the crystal. The incident light is therefore not doubly polarized but traverses the crystal with its original single vibration direction. As this vibration direction is perpendicular to a principal plane, the light is transmitted through the crystal with an ω index of refraction. Therefore, along both the EW and NS vertical sections, the incident EW plane-polarized light is parallel to a privileged direction within the crystal and is transmitted through the crystal unaltered as EW plane-polarized light.

Incident plane-polarized EW light that enters the crystal along any other vertical section (such as the NE–SW plane in part D) has a vibration direction that is neither parallel nor perpendicular to that section. Consequently, this plane-polarized light is resolved into the two privileged vibration directions permitted by the crystal. The vibrational component perpendicular to the section has an ω index of refraction; the vibrational component parallel to the section has an ε' index of refraction.

A composite picture of the various vibration directions is shown in Figure 6-2. Along the NS and EW vertical sections, plane-polarized light is retained. This light has an ε' index in the EW plane and an ω index in the NS plane. At any other point within the four quadrants, two mutually

Figure 6-1
(A) A uniaxial grain is oriented with its c axis perpendicular to the stage, and the condensing lens is inserted. Traces of vertical sections are indicated as NS, EW, and NE–SW. The NE–SW section represents any randomly chosen section that is not EW or NS. (B) The vertical EW section. (C) The vertical NS section. (D) The vertical NE–SW section. The vibrational patterns within the grain are shown beneath each of the vertical sections. Refraction is ignored here and in Figure 6-3. See text for explanation.

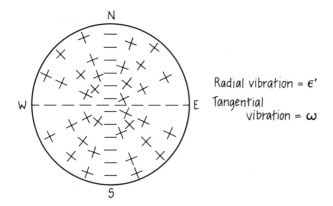

Figure 6-2
A composite sketch of the various vibration directions within the grain described in Figure 6-1. A single vibration direction is present only in the NS and EW vertical planes.

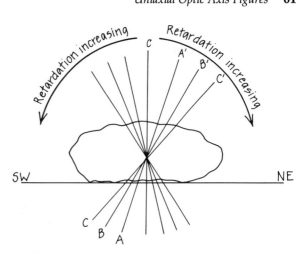

Figure 6-3
The NE–SW section of Figure 6-1. The ray parallel to the grain's c axis has a single index, ω, and undergoes no retardation. Retardation increases as the ray direction deviates from this orientation.

perpendicular vibrations are developed. One of these is *radial*, or parallel to a plane that includes the c axis (and has an ε' index), while the other is *tangential*, or perpendicular to such a plane, and has the index ω.

Upon leaving the top of the crystal, mutually perpendicular vibrations combine into a single vibration. Depending upon the grain thickness, the vertical angle of the ray path, and velocity differences, various amounts of retardation occur. As seen earlier, different amounts of retardation produce different interference colors (Plate 1). Thus each of the four quadrants will contain a variety of interference colors. Along the NS and EW sections, however, only a single vibration is present, so no retardation is developed. These single vibrations are eliminated at the upper polarizer. The interference figure that is produced thereby consists of a black cross with arms that trend NS and EW. Within the quadrants, one or more interference colors are present.

The black bands that make up the arms of the cross are called *isogyres*. The point where the arms meet marks the position of the c axis and is called a *melatope* (literally, black spot). Variously colored curves that may be found within the four quadrants are called either *isochromes* or *isochromatic curves*. The isochromes form a concentric pattern about the melatope: the farther from it, the higher the colors. The pattern is due to the increasing amounts of retardation at increasing angular distance from the c axis, as well as to the increase in path length at greater angles. If, however, the difference between the ε and ω indices of refraction of the mineral is small (low birefringence), only white or gray interference colors will be seen in the interference figure.

The sequence of interference colors is understood when we consider again the NE–SW vertical section (Fig. 6-3). The one light ray parallel to the c axis of the crystal undergoes no retardation. Along the ray direction AA', slightly inclined from the c axis, both ω- and ε'-type vibrations are developed. The ω-type vibration has a fixed index of refraction (and velocity). The ε'-type vibration has an intermediate index between ω and ε; the closer the ray direction is to that of the c axis, the closer ε' is to ω. (Recall that ε' values approach

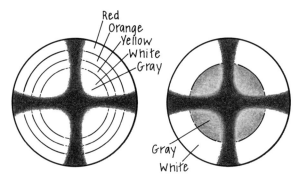

Figure 6-4
Uniaxial optic axis interference figures. The interference colors present in the figure on the left are of higher levels than those on the right; assuming identical grain thicknesses, the higher levels of interference colors are indicative of greater birefringence.

ε when the ray direction becomes more nearly perpendicular to the c axis.) It follows that for ray AA', the velocity difference between the two vibrations is small, and consequently the retardation developed for this ray direction is also small; thus the interference color developed from ray AA' is relatively low.

The ray direction BB' is at a greater angle to the c axis, and the value of ε' for this ray is farther from ω for ray BB' than for ray AA'. It follows that the retardation and level of interference colors is also greater. Ray CC', in turn, produces still higher levels of interference colors. The retardation developed by highly inclined rays such as CC' is also enhanced by an increase in ray path within the crystal. The generality here is that the interference colors produced in this section are higher as the ray direction diverges from parallelism with the c axis.

The above rule follows equally well for *any* (non-EW or -NS) section. The interference colors found in the interference figure are directly related to angular distance from the melatope, but they are influenced by grain thickness and birefringence as well. Figure 6-4 shows two uniaxial optic axis figures. (See also Figures 1 and 2 in Plate 2.) Both have isogyres that intersect at the melatope. Both also have isochromatic curves concentric to the melatope, but in different numbers. The number of isochromes present is a function of both grain thickness and birefringence. Transmission through a thicker grain causes greater retardation than a thinner one. Also, because birefringence is simply the difference between the maximum and minimum indices of refraction (ε and ω), minerals with high birefringence produce higher interference colors than equally thick minerals with lower birefringence. Consequently when working with several minerals of equivalent thickness (as one does when examining thin sections), it should be kept in mind that the number of isochromatic curves directly reflects the birefringence of the various minerals.

UNIAXIAL INCLINED OPTIC AXIS FIGURES

Many grains yield uniaxial interference figures in which the c axis is inclined (and not perpendicular) to the microscope stage. Such orientations are favored by the fairly common tendency of uniaxial minerals to possess pyramidal, rhombohedral, or scalenohedral cleavages. It is almost impossible, for example, to find fragments of the hexagonal carbonates in which the c axes are not inclined to the stage, because of their almost perfect rhombohedral cleavage. Grains in this orientation can be identified as such because they do not possess the maximum interference colors consistent with their thickness and birefringence.

Inclination of the c axis results in inclination of the entire vibration pattern (Fig. 6-5); inclination of the vibration pattern results in a similar inclination of the interference figure. In determining the optic sign from inclined uniaxial interference figures (see p. 72), it is necessary to know which part of the interference figure is being observed.

The isogyres of an optic axis interference figure divide the figure into four equal portions. The four quadrants are conveniently referred to as NE, NW, SE, and SW (Fig. 6-6). There is no problem in determining which quadrants are being observed when the melatopes are in the field of

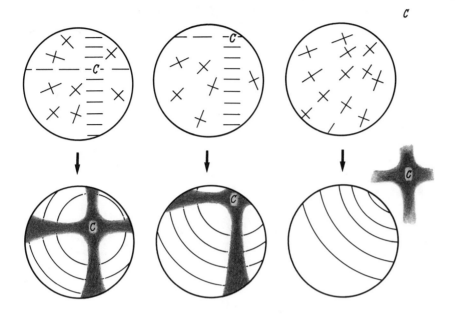

Figure 6-5
Uniaxial inclined optic axis interference figures. Inclination of the c axis of the grain (from an initial position of being normal to the stage) causes inclination of the entire vibrational pattern and hence of the interference figure.

view. When the melatopes are outside of the field of view, their positions can be deduced by observation of the isochromes or the isogyres.

Figure 6-7 shows an inclined optic axis figure in which the c axis of the crystal is inclined to the NE. The grain orientation is easily determined because the isogyres intersect to the NE (out of the field), and the concentric isochromes have their common center to the NE. The quadrant observed is SW. In Figure 6-8, the common center (the c axis) of the concentric isochromes is to the east of the field of view. Parts of the NW and SW quadrants are observed. Figure 6-9 shows only isochromes. It is obvious, however, that the SW quadrant is being observed, because the common center of the isochromes (the c axis) lies to the NE.

Determination of an unknown observed quadrant is slightly more complex when isochromes are not distinct and only a single isogyre is present. It becomes necessary to rotate the stage to determine the position of the melatope and, in turn, the quadrant observed. Figure 6-10 shows a sequence of isogyre positions that occur during rotation of the stage. The melatope remains at a

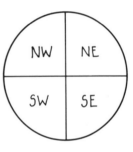

Figure 6-6
The four quadrants of a uniaxial optic axis figure are named as indicated.

Figure 6-7
The melatope is located to the NE. The SW quadrant is seen.

Figure 6-8
The melatope is to the east, and the NW and SW quadrants are seen.

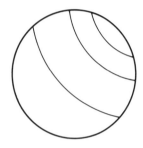

Figure 6-9
The melatope is inclined to the NE, and the SW quadrant is seen.

fixed angle from the center of the field of view during rotation of the stage, and the isogyres retain their NS and EW orientations during rotation.

One way of determining the particular quadrant that is observed is shown in Figure 6-11. First rotate the stage until no isogyres are present in the field (part A). Then rotate the stage both clockwise and counterclockwise until isogyres enter the field; the extrapolated intersection of these isogyres marks the position of the melatope. In part B, a clockwise rotation has brought an EW-oriented isogyre in from the north; a counterclockwise rotation (part C) brings a NS-oriented isogyre in from the east. Knowing that there are isogyres to the north and the east (part D), it is easy to see that their intersection point is to the NE and that the quadrant being observed is the SW.

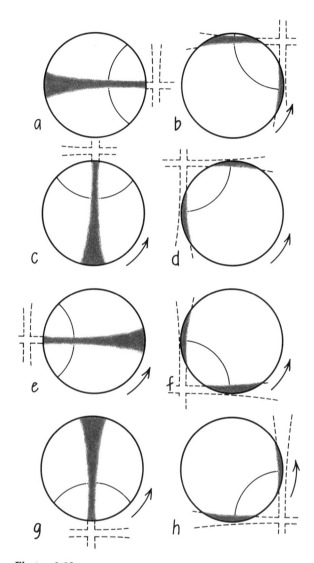

Figure 6-10
A uniaxial inclined optic axis figure. The melatope is just outside of the field of view. Counterclockwise stage rotations (at 45° intervals) are shown in the sequence (A) through (H). Notice that the isogyres change in position, but not in orientation. At high inclinations of the c axis, the distal ends of the isogyres curve slightly when entering and leaving the field of view.

Uniaxial Optic Normal Figures

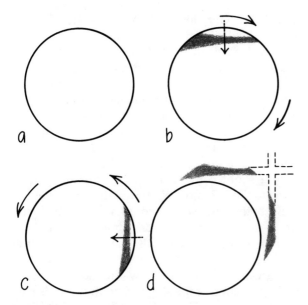

Figure 6-11
A procedure for finding the location of the melatope in a uniaxial inclined optic axis figure. (A) The problem: no isogyres or isochromes are visible in the field of view. (B) A small clockwise stage rotation brings an isogyre into the field. In this example the isogyre extends EW. The melatope must be to the NE or NW. (C) A small counterclockwise rotation removes the first isogyre. In this example, a second isogyre enters the field from the east. (D) The intersection point of the two isogyres that were seen in (B) and (C) must lie to the NE, hence the view is of the SW quadrant.

UNIAXIAL OPTIC NORMAL FIGURES

Optic normal figures are so named because the observer views the figure normal to the optic axis when the c axis of the crystal is parallel to the stage. This type of figure is also called a *flash figure*, because the isogyres move rapidly into and out of the field during a small rotation of the stage. Grains in this orientation possess a maximum interference color that is consistent with their thickness and birefringence.

The origin of the optic normal figure can be understood with the aid of Figure 6-12. Part A

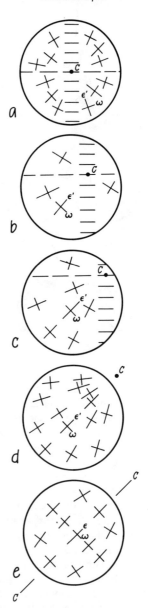

Figure 6-12
Vibrational patterns that yield uniaxial interference figures. In (A) the c axis of the grain is perpendicular to the stage. The sequence (B) through (E) shows the grain's c axis becoming parallel to the stage. Notice that with increasing parallelism of the c axis to the stage, each type of vibration direction in the field of view approaches parallelism. In (E), the ε and ε' vibration directions are essentially NE–SW and the ω vibration directions are essentially NW–SE.

66 Chapter 6: Uniaxial Interference Figures

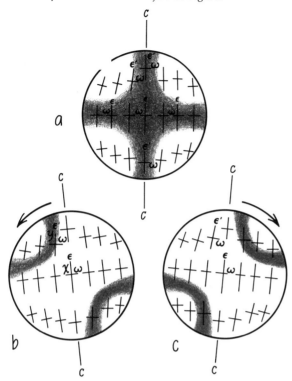

Figure 6-13
Behavior of the optic normal interference figure. The permissible vibration directions that are parallel (or inclined) to the c axis have the ε (or ε') index of refraction; vibrations normal or almost normal to the c axis have the index ω. (A) The c axis is oriented NS. Within the shaded area, permissible vibration directions are essentially NS and EW. As the EW permissible vibration direction is parallel to the plane-polarized light from the lower polarizer, this light retains its single plane of polarization and is eliminated by the upper polarizer. The resulting interference figure contains a diffuse isogyre cross. Permissible vibration directions in the four quadrants are a few degrees away from NS and EW, and the incident plane-polarized light is doubly polarized. Some of this light, transmitted through the upper polarizer, is seen as color. (B) Counterclockwise rotation of the stage: in the NW and SE quadrants, some of the permissible vibration directions (such as at point Y) become parallel to the plane-polarized EW light and are seen as isogyres (shaded), while areas that were in isogyres in part (A) (as at point X) rotate out of extinction and thus transmit some light through the upper polarizer to produce color. (C) A clockwise rotation of the stage results in isogyres moving into the NE and SW quadrants.

shows the vibration pattern when the c axis is perpendicular to the microscope stage. The sequence from B through E progressively inclines the c axis until it becomes parallel to the stage (NE–SW). In this sequence there is a gradual increase in the degree of parallelism of both the ε' and ω types of vibrations, just as lines of latitude and longitude become more parallel as one moves from the pole of a globe to the equator. In part E, where the c axis is parallel to the stage, there is only a slight convergence of ε' vibration directions to the NE and SW. Across the entire field of view, both ε' and ω vibrations have essentially the same orientations. The orientation of these vibration directions controls the position of the isogyres.

We can now deduce how the isogyres move during rotation of the stage when the c axis is parallel to the stage. First consider Figure 6-13A; here the c axis of the crystal is oriented NS. Permissible vibration directions are shown as usual, with ε and ε' vibration directions essentially NS and ω vibration directions essentially EW. Note that these are the *possible* vibration directions. If the incident plane-polarized light coincides with one of the two directions, the light has no component in the second direction, and the second vibration will not occur.

The permissible vibration directions in the darkened area are oriented exactly or almost exactly NS and EW. The convergent EW plane-polarized light entering the crystal from the condensing lens below coincides with the permissible EW vibration direction in the crystal. Within the shaded area, therefore, the incident light does not become doubly polarized but is transmitted through the crystal unchanged, only to be eliminated by the upper polarizer. The shaded part of the field is seen as a diffuse black cross.

In the unshaded outer quadrants of the field, the permissible vibration directions do not coincide with the incident plane-polarized EW light. In these areas the incident light is doubly polarized, and some gets through the upper polarizer.

Thus, in this orientation the interference figure is a large, diffuse central cross with minor amounts of light and color to the NE, NW, SE, and SW. In many optic normal figures, the pattern is extremely diffuse, making it difficult even to distinguish among light and dark areas.

If the stage is rotated slightly in a counterclockwise or clockwise direction (Fig. 6-13B and C), the black cross breaks into two segments that leave the field of view. Rotation of the stage changes the orientation of the permissible vibration directions of the crystal, which in turn results in changing the location of the isogyres. Consider point X in part B. Due to the slight rotation of the stage, the possible vibration directions at point X are no longer EW and NS; the incident EW plane-polarized light becomes doubly polarized, and the isogyre leaves this point. However, at a point such as Y, the permissible vibration directions are now NS and EW, and the incident light remains plane-polarized EW. This light is eliminated by the upper polarizer, which means that point Y lies in an isogyre. Isogyre locations similar to Y are indicated in the shaded areas of the diagram. Further rotation removes the isogyres from the field of view, as none of the possible vibration directions are parallel to the incident beam. The rapidity of the change in position of the isogyres is why the figure is called a flash figure.

The c axis is oriented NS and parallel to the stage in Figure 6-13A. A counterclockwise rotation (part B) causes the c axis to move toward a NW–SE orientation; the isogyres leave the field of view in the NW and SE quadrants. A clockwise rotation (part C) from the orientation in part A causes the c axis to move toward the NE–SW, and the isogyres leave the field of view in the NE and SW quadrants. Knowing this relationship, it is easy to locate the c axis using a flash figure. Rotate the stage until a black cross is present; then rotate the stage slightly in either direction until the isogyres begin to leave the field. The c axis of the crystal lies in the quadrants within which the isogyres leave the field.

Uniaxial optic normal figures grade into highly

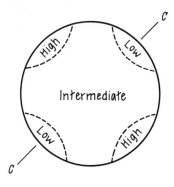

Figure 6-14
When the c axis is oriented diagonally in an optic normal interference figure, no isogyres are present. The quadrants in which the c axis lies correspond to the lowest level of interference colors; colors are higher in the center of the field of view and reach maximum values in those quadrants not occupied by the c axis. Generally this pattern cannot be observed because of diffuseness of the higher order colors.

inclined optic axis figures when the c axis of the crystal is not exactly parallel to the stage. The way to decide between the two figures is to measure the number of degrees of stage rotation within which the isogyres are in the field of view. If the isogyres enter and leave within a stage rotation of 15° or less, the figure is an optic normal type. Greater values indicate a highly inclined optic axis figure.

The interference colors seen in an optic normal interference figure are generally diffuse and difficult to evaluate. The colors are best observed when the c axis of the crystal is in the 45° position (diagonal), as in figure 6-14. By analogy with the uniaxial optic axis figure (Fig. 6-4), it is reasonable to infer that the interference colors in the quadrants containing the c axis (NE and SW) are those of the lowest level. Somewhat higher level colors occur in the center of the field, and maximum level colors are found in the opposite (NW and SE) quadrants. When the colors are distinct enough to be differentiated, the location of the c axis is known.

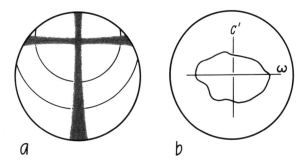

Figure 6-15
(A) A uniaxial inclined optic axis figure: the location of the melatope (to the north) indicates that the c axis lies in the NS vertical plane. (B) The same grain viewed orthoscopically: the inclined c axis (shown as c') is in the NS vertical plane. The vibration direction for ω is therefore EW, parallel to the permissible vibration direction of the lower polarizer. A Becke test on this grain gives an estimate of the ω index.

USES OF UNIAXIAL INTERFERENCE FIGURES

Interference figures are used to distinguish between uniaxial and biaxial materials, to reveal crystal orientation, and to ease determination of indices of refraction and optic sign.

Determination of Anisotropic Character

Uniaxial highly inclined optic axis and optic normal figures can be confused with similar-appearing biaxial figures; however, well-centered optic axis figures serve to distinguish uniaxial from biaxial materials. If the unknown material is uniaxial, the isogyre cross is maintained during stage rotation. If the material is biaxial, two melatopes are present (see Fig. 8-24), and rotation of the stage results in movement of the isogyres to produce a cross in some positions, and two separate isogyres in others.

The technique for determining uniaxial versus biaxial character is to find an interference figure in which at least one melatope lies within the field of view. Such figures are obtained from grains with relatively low interference colors.

Crystal Orientation and Index Measurement

The position of the melatope indicates the orientation of the c axis of the crystal. If the melatope coincides with the cross-hair intersection, the c axis is perpendicular to the stage. Such a grain permits a Becke line determination of ω. If the melatope does not coincide with the cross-hair intersection but is either within or relatively close to the field of view, the figure is of the inclined optic axis type. Grains of this orientation permit a measurement of ω and ε'.

In order to measure ω, rotate the stage until the melatope lies along the NS cross hair (Fig. 6-15A). This indicates that the inclined c axis is within the vertical NS plane. Viewed orthoscopically under crossed nicols, the grain should now be in an extinction position (Fig. 6-15B) or a few degrees from extinction. After bringing the grain to extinction, uncrossing the nicols will permit a measurement of the ω index. To make a Becke test of ε', the procedure is the same except that the melatope is rotated to coincide with the EW cross hair before orthoscopic examination.

An optic normal figure indicates that the c axis of the crystal is parallel to the stage. Such a crystal permits a measurement of both ω and ε. This orientation is necessary for a true measurement of ε rather than ε'.

To measure ε, rotate the stage until the isogyres leave the field of view. The quadrants in which the isogyres leave are also the quadrants in which the c axis of the crystal lies. Assume that the isogyres leave the field in the NE and SW quadrants (Fig. 6-16A). Convert the microscope to orthoscopic illumination: the grain displays interference colors (Fig. 6-16B). Rotate the stage clockwise until the crystal goes into extinction (Fig. 6-16C), which places the c axis in an EW horizontal position. As the ε vibration direction coincides with the c axis, it is now possible to uncross the nicols and perform a Becke test for ε. Conversely, ω can be measured by putting it in the EW position, rotate 45° counterclockwise to extinction from the position indicated in Figure 6-16B. By now it should be clear how important it is that

Uses of Uniaxial Interference Figures

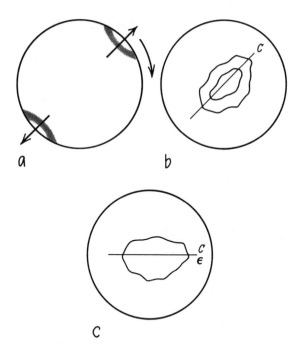

Figure 6-16
The measurement of ε with the aid of a uniaxial optic normal figure. (A) After obtaining an optic normal figure, rotate the stage until the isogyres leave the field in the NE and SW quadrants. Continue the rotation about 45° from the isogyre cross position. The c axis is now oriented roughly NE–SW. (B) Viewed orthoscopically the grain now shows interference colors. (C) Rotate the stage clockwise (about 45°) until the grain is at extinction. The c axis and the ε vibration direction are now EW. Uncross the nicols and perform the Becke test for ε.

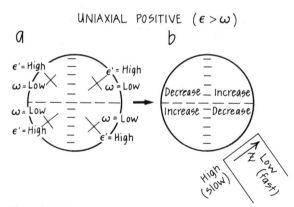

Figure 6-17
(A) The vibration directions that produce a uniaxial optic axis figure. Along the NS and EW cross hairs there is a single EW vibration direction (parallel to the permissible vibration direction of the lower polarizer). Within the four quadrants there are, at each point, two vibration directions; a radial vibration is always an ε' type, and a tangential vibration is always ω. (B) For a *positive* uniaxial mineral, an accessory inserted in standard orientation (high-vibration direction NE–SW) increases retardation in the NE and SW quadrants and decreases retardation in the NW and SE quadrants. The region originally occupied by the isogyres assumes the interference color of the accessory.

you have determined the privileged direction of the lower polarizer in your microscope correctly! Knowing ε and ω permits, in turn, the determination of birefringence—the numerical difference between ε and ω. Many determinative tables of minerals are based on this value.

Determination of Sign

Finding the optic sign of a uniaxial crystal is simple if an interference figure is obtainable. The determination is a matter of inserting an accessory plate and observing changes of retardation in the various quadrants.

Figure 6-17A shows the vibration directions that produce a uniaxial optic axis figure. Observe once again that radial vibrations have an index ε' whereas tangential vibrations are ω. In this example, the mineral is uniaxial positive ($\varepsilon > \omega$). An accessory (in the standard orientation) is inserted from the SE. The high index direction of the accessory coincides with the high index direction (ε') of the interference figure in the NE and SW quadrants and opposes it in the NW and SE quadrants. Retardation is thus additive in the NE and SW quadrants and subtractive in the NW and SE quadrants.

An increase in retardation due to addition results in an increase in the level of the interference colors of the figure; a decrease in retardation due to subtraction lowers the colors. In Figure 6-18,

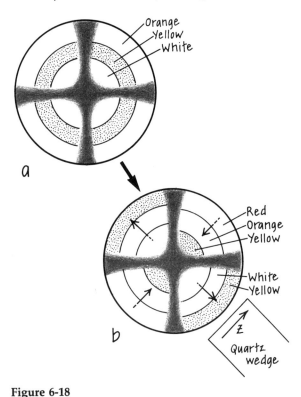

Figure 6-18
(A) A *positive* uniaxial optic axis figure showing several isochromatic curves. When isochromes are present, the quartz wedge is generally preferred for sign determination. (B) During insertion of the quartz wedge, increase in retardation in the NE and SW quadrants is seen as an inward movement of isochromes; decrease in retardation in the NW and SE quadrants is seen as an outward movement of isochromes.

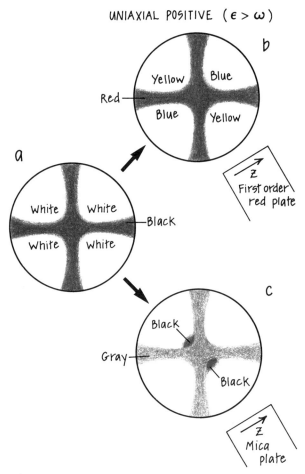

Figure 6-19
(A) A *positive* uniaxial optic axis figure. Isochromatic curves are not present and the quadrants are mainly white; thus the first-order red plate or mica plate is preferred for sign determination. (B) The first-order red plate produces blue to blue-green in the NE and SW quadrants (by addition) and yellow in the NW and SE quadrants (by subtraction). (C) Insertion of the mica plate produces black spots adjacent to the melatope in the NW and SE quadrants through subtraction to zero retardation. The slight increase in retardation in the NE and SW quadrants will show no obvious color changes except perhaps for the appearance of some yellow at the edge of the field of view.

several isochromes are present and the preferred accessory is the quartz wedge. Continuous insertion of the quartz wedge continuously increases interference colors in the NE and SW quadrants (Fig. 6-18B). This is seen as an inward migration of isochromes, with higher level colors replacing lower level colors. In the NW and SE quadrants, a decrease in retardation is seen as an outward flow of isochromes.

In a crystal of low birefringence (Fig. 6-19), the four quadrants of the uniaxial optic axis figure are essentially white. Assume again a positive optic sign. With insertion of the first-order red plate

(Fig. 6-19B), interference colors increase in the NE and SW quadrants (white plus the red of the first-order plate yields blue or blue-green). In the NW and SE quadrants, subtraction occurs (yielding a first-order yellow below the first-order red). See also Figures 1 and 2 in Plate 2.

The mica plate is often used instead of the first-order red plate to determine the optic sign on interference figures with few or no isochromes. Figure 6-19C shows the color reaction for a positive uniaxial mineral. In quadrants in which the birefringence is additive, there is a slight inward movement of isochromes. In the subtractive quadrants, two black dots appear adjacent to the melatope (gray of the mica plate compensates the gray of the interference figure).

In summary, the effect of either accessory on a positive uniaxial optic axis figure is to produce an additive retardation in the NE and SW quadrants. For figures displaying isochromes, a quartz wedge is preferred, and for figures containing no isochromes either the first-order red or the mica plate is preferred.

Figure 6-20 shows the situation for a uniaxial

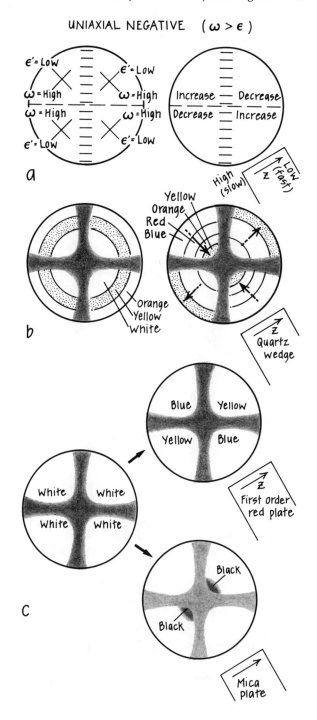

Figure 6-20
(A) For a *negative* uniaxial mineral, an accessory decreases retardation in the NE and SW quadrants and increases retardation in the NW and SE quadrants. The region originally occupied by the isogyres assumes the interference color of the accessory. (B) A *negative* uniaxial optic axis figure with isochromes; the quartz wedge is preferred for sign determination. During insertion of the quartz wedge, isochromes move outward in the NE and SW quadrants and inward in the NW and SE quadrants. (C) A *negative* uniaxial optic axis figure without isochromes. Either the first-order red plate or mica plate is preferred for sign determination. The red plate decreases retardation to yellow in the NE and SW quadrants and increases it to blue or blue-green in the NW and SE quadrants. Insertion of the mica plate results in the formation of black spots adjacent to the melatope in the NE and SW quadrants. A slight increase in interference color in the NW and SE quadrants is usually not apparent, but some yellow may appear at the edge of the field.

72 Chapter 6: Uniaxial Interference Figures

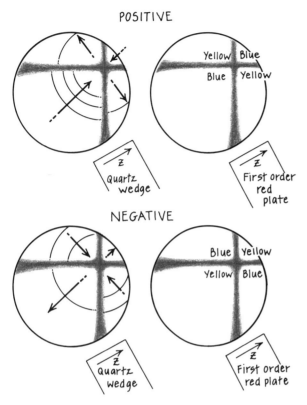

Figure 6-21
Inclined optic axis figures produce the same sign reactions as centered optic axis figures.

negative mineral: tangential vibrations, again, have the index ω and the radial vibrations have an ε' index. As the mineral is uniaxial negative, ε' is the lower index and ω the higher ($\omega > \varepsilon$). With insertion of an accessory, retardation is additive in the NW and SE quadrants and subtractive in the NE and SW quadrants. The result is that isochromes flow inward in the NW and SE quadrants during quartz wedge insertion, and outward in the NE and SW quadrants (part B). Similarly, insertion of a first-order red plate over quadrants that are white produces blue (by addition) in the NW and SE quadrants and yellow (by subtraction) in the NE and SW quadrants (part C). The mica plate reaction is characterized by black spots adjacent to the melatope in the NE and SW quadrants.

The sign is equally easy to obtain on inclined optic axis interference figures. First determine the quadrant or quadrants that are being observed (see p. 64). Insertion of an accessory produces the same type of interference color changes as seen in an optic axis figure (Fig. 6-21). In the few cases where interpretation is difficult due to indistinct color effects, observe two quadrants simultaneously, so that a direct color comparison between them can be made.

Sign determinations can be made using uniaxial optic normal figures, either directly or indirectly. After an optic normal figure has been obtained, rotate the stage until the c axis of the crystal is in a diagonal position; in Figure 6-22A the c axis has been placed NE–SW. If the interference colors are sufficiently distinct, it will then be obvious that the colors are lower in the NE and SW quadrants and higher in the NW and SE quadrants. As the vibration directions that produce these colors are in approximately the same orientation throughout the entire field (see Fig. 6-13), inserting an accessory will cause the interference colors in the whole field of view to either all increase simultaneously or all decrease simultaneously.

If the mineral is positive ($\varepsilon > \omega$), the ε index is high and coincides with the high index direction of the accessory (Fig. 6-22B). This results in an increase in retardation. The effect seen during insertion of the quartz wedge is higher order colors flowing over lower order colors. If the mineral is of negative sign (part C), the flow of colors is the reverse.

In most cases, however, the interference colors of optic normal figures are too diffuse for adequate sign determination. There is no need to despair as the sign can be determined by an indirect technique. Assume again that an optic normal figure has been obtained. Rotation of the stage has placed the c axis in the NE–SW horizontal orientation, and the interference colors are the usual hopeless blur. Now change from conoscopic to orthoscopic illumination by removing

Uses of Uniaxial Interference Figures 73

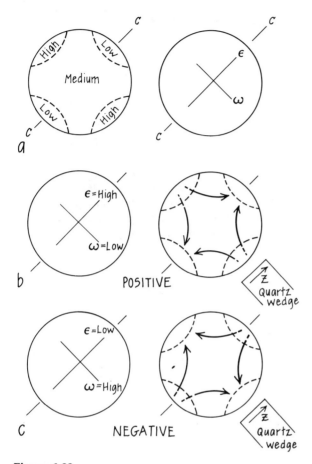

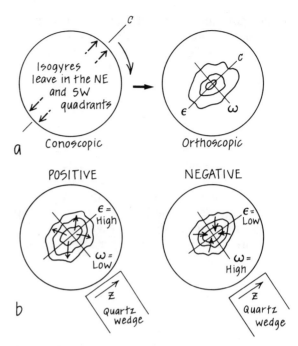

Figure 6-22
(A) The interference color pattern in a uniaxial optic normal figure; the c axis is oriented NE–SW. The interference colors are lowest within the quadrants occupied by the c axis. The ε vibration direction, as always, is parallel to the c axis and the ω direction is perpendicular to the c axis. (B) In a positive mineral, the high index direction (ε) is NE–SW and the low index direction (ω) is NW–SE. During insertion of a quartz wedge, interference colors flow from the NW and SE quadrants toward the center and from the center into the NE and SW quadrants. All colors simultaneously increase. (C) In a negative mineral, a quartz wedge causes interference colors to flow from NE and SW quadrants to the center and from the center to the NW and SE quadrants. All colors decrease simultaneously.

Figure 6-23
When uniaxial optic normal figures display only diffuse interference colors that do not yield a distinctive sign reaction, the sign can be determined indirectly. (A) After obtaining an optic normal figure, rotate the stage until the isogyres leave the field in the NE and SW quadrants. The c axis is now within the NE and SW quadrants. Convert to orthoscopic illumination. The fragment shows interference colors when the nicols are crossed. (B) Insert a quartz wedge and observe the movement of isochromes on the grain. An outward movement indicates a positive sign ($\varepsilon > \omega$), an inward movement a negative sign ($\varepsilon < \omega$).

the Bertrand and condensing lenses. The grain, being in a diagonal orientation, displays interference colors (Fig. 6-23A). The interference figure has revealed that the c axis of the grain is horizontal and NE–SW; from this it follows that the ε vibrations are also in the NE–SW direction, and the ω vibrations are oriented NW–SE. Insertion of a quartz wedge indicates which index is higher. Figure 6-23B shows that if the mineral is positive, the color bands move outward during insertion of the wedge. Inward movement of color bands indicates a negative sign. This technique can also be used for sign determination with an inclined optic axis interference figure in which the color effects are too blurred for proper interpretation.

ADDITIONAL READINGS

Becke, F. 1905. Die Skiodromen. *Tschermaks Mineralogisches und Petrographische Mitteilungen 24*, 1–34.

Fletcher, L. 1891. The optical indicatrix and the transmission of light in crystals. *Mineralogical Magazine 9*, 278.

Phemister, T. C. 1954. Fletcher's indicatrix and the electromagnetic theory of light. *American Mineralogist 39*, 172–192.

Wahlstrom, E. E. 1979. *Optical Crystallography*, 5th ed. New York, John Wiley & Sons, 242–268.

Wright, F. E. 1923. The index ellipsoid (optical indicatrix) in petrographic microscope works. *American Journal of Science 185*, 133–138.

7

Microscopic Identification of Unknown Uniaxial Materials

> It is not worth while to go round the world to count the cats in Zanzibar.
>
> HENRY DAVID THOREAU

Because identification of unknown uniaxial materials by oil immersion differs somewhat from identification of materials in thin section, the two procedures will be treated separately. It should be understood that materials identified in thin section are usually limited to either the common rock-forming minerals or materials within a compositionally restricted system, whereas immersion techniques may involve materials of almost any composition. This difference reflects the microscopist's ability to determine precise indices of refraction with the immersion technique; only approximate indices can be determined in most thin-section studies. If a material cannot be identified in the standard covered thin section, in some cases it can be extracted from the hand specimen and identified by immersion techniques.

Intermediate between standard thin-section and immersion techniques is the use of polished thin sections. The upper surface of the mounted rock slice is polished and left exposed, with no cover glass. This permits the use of reflected-light microscopy with opaque minerals (a specialized topic not taken up here) as well as polarized-light microscopy with transparent minerals. In addition, index measurements can be made on individual grains by removing some of the surrounding material (either mineral grains or mounting medium) and replacing it with an immersion liquid.

If the goal is simply mineral identification, it may not be necessary to determine all measurable parameters precisely, which could take a considerable expenditure of time. Obtain only the amount of information necessary to identify the material unambiguously. This procedure takes only a few minutes for most uniaxial materials.

OIL-IMMERSION TECHNIQUE

Before getting involved with the polarizing microscope, it is often a good idea to first examine a nonpulverized sample with a hand lens and binocular microscope. If only a pulverized sample is available, it is possible to determine the approximate hardness, and the chemical reaction with dilute hydrochloric acid. To get a rough idea of the hardness, sprinkle a few grains on the microscope slide. Put a second slide on top of the grains and slide one surface across the other. Remove the grains and examine the slide surface in reflected light with a hand lens. If the slide now has hairline scratches, the grains have a hardness greater than 5½ (Mohs scale).

As a matter of routine, apply a drop of dilute hydrochloric acid to a few grains on a glass slide. Effervescence indicates that the unknown material is a carbonate. A second possibility is that the grains may partially or completely dissolve in the acid. Such information is listed in most mineral descriptions. Use of other solvents (such as water) may furnish useful information as well. Naturally, keep chemical reagents away from good-quality optical equipment.

Make a standard grain mount as described earlier. The choice of immersion liquid is optional. Strongly colored materials commonly have higher indices of refraction than white or colorless substances (due to the presence of iron or other heavy cations), so it is usually best to choose an index liquid of about 1.65–1.67 for colored materials and one of about 1.57–1.59 for white or colorless materials.

It is best first to verify the uniaxial character of the substance under examination. Using the low- or medium-power objective and crossed nicols, scan the fragments and locate one or more that are characterized by relatively low interference colors and an average or greater than average grain size. Such grains have their optic axes perpendicular or near perpendicular to the stage. Switch to the high-power objective and conoscopic observation in order to obtain an interference figure. The melatope will usually be in the field of view; if not, try another grain with low interference colors. Rotate the stage and verify that the cross does not break into two separate isogyres. This confirms the uniaxial character of the material. Determine the optic sign with the appropriate accessory; use the quartz wedge if isochromes are visible, or the first-order red or mica plate if the field of view is white or gray.

Next, estimate the index ω. This may require a few additional mounts with appropriately chosen immersion liquids. Once the sign is known, any grain can be used for this purpose. Keep a record of the index liquid of each mount. It is usually not necessary to obtain a precise value. A determination to within 0.01 is often sufficient. During this procedure, also note whether the interference colors are low, medium (second order), or high. This information provides a rough measure of the birefringence.

At the same time it should also be noted whether any cleavage is present. Cleavage, as noted earlier, can be detected by either a constancy of interference colors within grains (indicative of a consistent thickness) or by the presence of straight grain edges. If grain edges are straight, the type of extinction should be noted, as well as the angular relationship of cleavage directions to the ω and ε index directions. Finally, a relatively consistent type of interference figure indicates the presence of a cleavage.

It may be useful to determine the specific gravity of an unknown material. Approximations may be made by noting whether the fragments float or sink within the immersion liquid. By adding a few coarse glass fragments to the standard mount, the cover glass is raised so that it can be determined easily, by precise focusing, whether the unknown material has floated up to the cover slip or has sunk down to the slide. Observations can be correlated with the specific gravities of the various immersion liquids. Figure 7-1 shows how the commonly used mixtures of α-monochloronaphthalene and diiodomethane vary in specific gravity as a function of index of refraction and temperature.

If the mineral possesses a color, this should be noted. Many colored minerals are pleochroic. If this is the case, note the colors associated with both the ω and ε vibration directions, and determine the pleochroic or absorption formula. The procedure is described on pp. 56–57. Recall that these colors are observed with uncrossed nicols.

After obtaining the above information, it is often possible to check determinative tables and identify the material, or perhaps obtain two or three reasonable possibilities.

If the mineral is not yet unambiguously identified, it is necessary to acquire additional information. Study the descriptions of the candidate minerals to determine what information is necessary. It may be that in one case a precise determination of ω is necessary. In another it may be necessary to determine ε. In still another, a particular angular relationship may exist between different

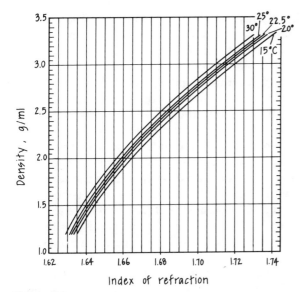

Figure 7-1.
The relationship between density, refractive index, and temperature for mixtures of diiodomethane and α-monochloronaphthalene. (Bloss, 1961. Reprinted by permission of CBS College Publishing.)

cleavages. Perhaps one mineral is commonly twinned and the others not. Choose the critical property and work on it. Constantly remember that if the primary purpose is mineral identification, the easiest and quickest way to do it is with minimum information.

THIN-SECTION TECHNIQUE

An unknown substance in thin section can seldom be identified with the aid of precisely determined indices of refraction, because the material to be identified is typically surrounded by other mineral grains or the cement that holds the section together. On the other hand, there are advantages gained in thin-section determinations: (1) the section (if made properly) is of a consistent thickness throughout (0.03 mm),[1] (2) the habit of the mineral is retained within the rock matrix, (3) mineral associations and incompatibilities may be determined by observation of other minerals present in the rock, and (4) the general character of the rock (igneous, metamorphic, or sedimentary) is commonly obvious from cursory examination.

First the uniaxial character of the mineral and its optic sign should be determined. Assuming that a number of grains are present in the section, use the low- or medium-power objective to find a grain with either relatively low interference colors or one in constant extinction; as all grains are of constant thickness, their size is of no importance here. An interference figure for a grain in constant extinction should have the melatope in the field of view; if it does not, the substance is either isotropic or of extremely low birefringence. Rotate the stage to verify that the isogyre cross remains intact, proving that the grain is indeed uniaxial. Determine the sign.

The approximate index of refraction is determined by considerations of relief and Becke tests. Find a grain at the edge of the section. Here the unknown substance is in contact with the section's cementing material; this is most likely epoxy ($n = 1.533-1.540$) in newer sections, or Canada balsam ($n = 1.537$) in older sections. The relief of the grain against the cement can be described (somewhat ambiguously) as low, medium, or high. Low relief is shown by minerals such as quartz and feldspar; moderate relief is shown by mica and chlorite; high relief is shown by olivine and garnet. The relief is enhanced by partial closure of the substage diaphragm. Check the Becke line movement between the unknown material and the cement (with partial closure of the substage diaphragm). This will reveal whether the unknown material has positive relief (higher n) or negative relief (lower n than the cement). This can also be done using adjacent known minerals within the section.

---

1. A wedge-shaped thin section is detected by a gradual rise or fall of interference colors as one moves across the section. Sections of nonstandard thickness are noted by a somewhat higher or lower maximum interference color than that expected in minerals of known and fixed birefringence. Quartz, for example, should be first-order white with a slight yellowish hue, as a consequence of a birefringence of 0.009. The Michel–Lévy chart is useful in determining a section's thickness (see Figure 7-2 and explanation).

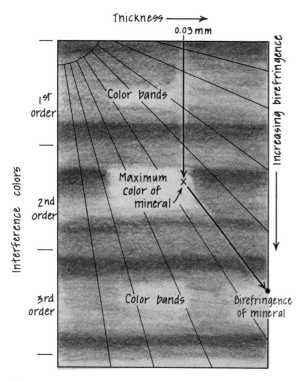

Figure 7-2.
A schematic view of the Michel–Lévy chart. One use of the chart is to determine birefringence of a mineral in thin section. Find the maximum interference color shown by the mineral in the thin section. Assuming that the section has been made to the proper thickness, 0.03 mm, proceed downward along the 0.03 mm thickness line to the level of the maximum interference color (point X in this example). From there proceed along or parallel to the diagonal line to the edge of the chart, where the birefringence of the mineral is indicated. A colored version of the chart is given in Plate 1. The above procedure can be reversed, starting with a mineral of known birefringence, in order to determine the thickness of a thin section.

Another useful measurement is that of birefringence—the difference between the maximum and minimum index. Scan the slide and find the grain of the unknown mineral that has the highest interference color. As all of the grains are of constant thickness in a good-quality thin section, the interference color of a particular mineral is related only to orientation and birefringence. A grain whose c axis is parallel to the stage exhibits the maximum interference color (as both the ω and ε vibration directions are parallel to the stage). After finding the grain with the maximum interference color, check the grain's orientation by means of an interference figure. The figure obtained should be of the optic normal type. If it is not, search the slide until a grain in this orientation is found. The optic normal figure is assurance that the c axis is parallel to the stage and that the maximum interference color is being observed. Obtain a Michel-Lévy interference color chart (Plate 1 and Fig. 7-2). Observe that the top of the chart is labeled with increasing thickness from left to right. Proceed down the vertical line indicating the thickness of the thin section (0.03 mm) as far as the maximum interference color shown by the unknown mineral. From this point proceed down along or parallel to the diagonal lines to the edge (right side or bottom) of the chart. The number printed just outside of the chart is the birefringence of the mineral. Common minerals with a similar birefringence are listed adjacent to the chart. (This method of determining birefringence cannot be used with immersion techniques because the mineral fragments are not of consistent thickness.)

Cleavage, if present, appears as parallel cracks within the mineral grains. Detection of cleavage is enhanced by partially closing the substage diaphragm. Determine the quality and number of cleavage directions. Recall that prismatic cleavages are parallel to the c axis (the ε direction) whereas basal pinacoid cleavage is perpendicular to the c axis. Extinction will be parallel to both of these cleavage types. Grains having pyramidal, rhombohedral, or scalenohedral cleavage (at an oblique angle to the c axis) become extinct symmetric to the cleavage cracks.

Crystal habit may be revealed by euhedral or subhedral grains (see Figs. 5-3 and 5-4). Color and pleochroism should be noted as well. The presence, type, or lack of twinning can also be useful in identification, as discussed below. In addition, examine photomicrographs of the common minerals, such as those in Volume 2 of this book.

Knowledge of petrography and petrology is also an aid in mineral identification. Dietrich and Skinner (1979), Ehlers and Blatt (1982), and Williams, Turner, and Gilbert (1982) have descriptions, discussions, and photomicrographs of the common minerals and their associations. In addition, determinative tables, such as those in Volume 2, furnish some of this type of information.

The professional petrographer seldom goes through the entire ritual just described; he or she can identify many minerals either by sight or perhaps with the determination of a single parameter. The secret behind this kind of performance is that the expert has seen these minerals in many other thin sections. If thin sections are made properly, the same mineral will look much the same in each section. The "look" of a mineral may be based on a combination of birefringence, relief, cleavage, habit, the presence or absence of twinning, characteristic alteration products, and associations. For this reason it is wise to examine a large variety of well-described thin sections of rocks from a variety of geological environments. After a short time of such study, many of the rock-forming minerals can be identified on sight, but one should check one or two properties to be sure. There is considerable truth in the idea that the best petrographers are the people who have seen the greatest number of thin sections.

ADDITIONAL FACTORS

Twins

When two or more crystals of the same species are intergrown in a special crystallographic relationship, the composite individual is known as a twinned crystal. Twins may be produced during crystal growth, through polymorphic inversion, or by deformation. Twins may be simple (two individuals), polysynthetic (more than two individuals according to the same law and on parallel composition planes), or cruciform (interpenetrating individuals that may form a cross). The feldspars are notable for their twinning behavior. The presence or absence of twins may be a useful aid in mineral identification.

Anomalous Interference Colors

A few minerals show anomalous interference colors: this may be either a slight deviation from the normal interference color sequence (for instance, the first-order white may have a bluish cast), or a distinct deviation from the normal color sequence. Some minerals possess a blue or brown interference color that does not vary as a function of grain thickness.

Anomalous interference colors of anisotropic materials result from either selective absorption of particular wavelengths, or non-parallelism of dispersion curves. When the dispersion curves are nonparallel, the birefringence varies as a function of the wavelength of the transmitted light. In some cases the dispersion curves cross, and the birefringence is zero for a particular wavelength. For that wavelength the substance is isotropic; a particular color is eliminated from the interference color sequence, and the remaining depleted sequence is anomalous. In other instances, the dispersion curves are nonparallel and nonintersecting. Here there may be particular combinations of grain thickness and birefringence that result in path differences of full wavelengths for particular wavelengths of light. That wavelength is eliminated by the upper polarizer, and an anomalous depleted interference color sequence results.

Variation in Index of Refraction

The indices of refraction measured for a particular mineral specimen may vary considerably from the particular values listed in reference books and determinative tables. This is a consequence of the fact that many mineral groups (such as tourmaline, apatite, and scapolite) have a wide composition range as a result of solid solution. The indices of refraction generally change gradually from one compositional end-member to another. Such minerals can usually be identified rather easily because other characteristics such as habit, birefringence, cleavage, or associations are distinctive.

ADDITIONAL READINGS

Deer, W. A., R. A. Howie, and J. Zussman. 1966. *An Introduction to the Rock-Forming Minerals*. New York: John Wiley & Sons, 528 pp.

Fleischer, M., R. E. Wilcox, and J. J. Matzko. 1984. *Microscopic Determination of the Nonopaque Minerals*. U.S. Geological Survey Bulletin 1627, 62–123.

Phillips, W. R., and D. T. Griffen. 1981. *Optical Mineralogy: The Nonopaque Minerals*. San Francisco: W. H. Freeman, 677 pp.

Troeger, W. E. 1979. *Optical Determination of the Rock-Forming Minerals*, 4th ed. (In English, by H. U. Bambauer, F. Taborszky, and H. D. Trochim). Stuttgart: Schweizerbartsche, 188 pp.

Winchell, A. N. 1951. *Elements of Optical Mineralogy, II, Descriptions of Minerals*, 4th ed. New York: John Wiley & Sons, 551 pp.

8

Biaxial Materials and Light

> No matter how bad things get you got to go on living, even if it kills you.
> — Sholom Aleichem

Most minerals are biaxial—orthorhombic, monoclinic, or triclinic—rather than uniaxial. Biaxial crystals have three unique crystallographic directions (a, b, and c), in contrast to the two of uniaxial materials (the equivalent a axes and c). It follows that biaxial materials also have three unique principal optical directions. These *principal optical directions* (X, Y, and Z) are always at right angles to each other (parallel to planes of maximum and minimum polarization), in contrast to the crystallographic axial directions, which may not be mutually perpendicular. Optical and crystallographic directions thus may or may not coincide. The planes formed by any pair of principal optical directions (XY, XZ, and YZ) are called *principal sections* (Fig. 8-1). Each of the three principal optical directions is associated with a unique vibration direction and index of refraction. A wave whose vibration direction is parallel to X corresponds to the smallest index of refraction, α; the intermediate index, β, corresponds to Y, and the maximum index, γ, corresponds to Z. The intermediate index β can have any value between that of α and γ. The three principal optical directions will be referred to as *principal index directions* in this book when attention is directed toward the associated index of refraction.

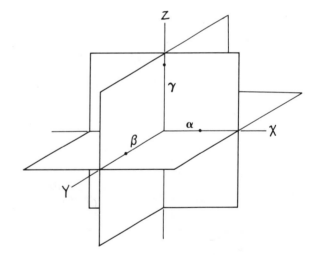

Figure 8-1.
Biaxial minerals have three principal indices: α, β, and γ. The vibration directions of these three indices are the lines X, Y, and Z respectively. The vibration directions are mutually perpendicular. Shown here are the three principal planes (XY, XZ, and YZ), each of which includes the vibration directions of two of the principal indices.

82   Chapter 8: Biaxial Materials and Light

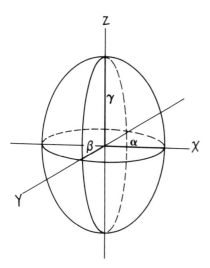

Figure 8-2.
A biaxial indicatrix. The principal radii correspond in length to the numerical values of the principal indices α, β, and γ. The three principal planes are shown where they meet the indicatrix.

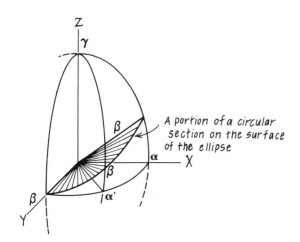

Figure 8-3.
A portion of a circular section (heavy line) within a biaxial indicatrix. As the circular section has a constant radius equal to β, it must be oriented so as to include the Y optical direction and be inclined between the XY and YZ principal planes.

INDICES OF REFRACTION AND THE INDICATRIX

The biaxial optical arrangement may be described conveniently in terms of an indicatrix—the geometric surface constructed to show vibration directions and indices of refraction (Fig. 8-2). Because of the symmetry of the orthorhombic, monoclinic, and triclinic systems, the biaxial indicatrix is a triaxial ellipsoid, that is, an ellipsoid that has three mutually perpendicular semiaxes, each of which is different in length. The positions of the three semiaxes give the directions of vibration (not transmission) that yield the three principal indices. The ellipsoid is constructed with the smallest radius proportional in length to the index α (in the X direction) and the largest radius proportional in length to the index γ (in the Z direction); the intermediate radius is proportional in length to the index β (in the Y direction). Other intermediate radii are called α' if their lengths are between those of α and β, and γ' if their lengths are between those of β and γ. Thus, the sequence of increasing radii is α, α', β, γ', γ.

Each of the principal sections of the indicatrix is elliptical. In fact, with the exception of two *circular sections*, every section through the center of a triaxial ellipsoid is elliptical. The two circular sections are of special interest and will be considered in detail. Let us first determine where they are located.

Consider the XZ plane in Figure 8-3; the shortest radius within this plane is α, and the longest is γ. Between these two extreme radii, there is an intermediate radius whose length is equivalent to the intermediate index β; this radius is indicated in Figure 8-3. Now find a radius α' on the XY principal plane. Between α' and γ there must be an intermediate radius equivalent in length to β; this, too, is indicated on the figure. Considering that there are an infinite number of α' radii that we could have chosen within the XY plane, there must also be an infinite number of points of radius β located between any chosen α' and γ. When these points are joined (see heavy line), a line of constant radius β can be drawn on the sur-

Indices of Refraction and the Indicatrix   83

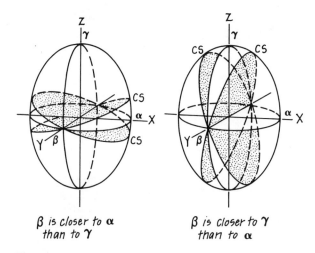

Figure 8-4.
The orientation of the two circular sections (cs) of a biaxial indicatrix is dependent upon the value of β as compared to α and γ. When β is close to α, the circular sections (left) make an acute angle to α. When β is close to γ, the circular sections (right) make an acute angle to γ.

face of the indicatrix. Additional consideration will show that this line can be extended completely around the indicatrix, forming a circle. Figure 8-4 shows that a second circle can be constructed in a symmetrically equivalent position. The two circular sections intersect on the Y coordinate (which, by definition, has the radius β).

Notice in the two diagrams of Figure 8-4 that the two pairs of circular sections have different inclinations to X and Z. This difference in inclination reflects the value of β relative to α and γ. If β is closer in value to α than it is to γ, the circular sections lie at a smaller angle to X than to Z. Alternatively, when the value of β is closer to γ than α, the circular sections lie at a smaller angle to Z than to X.

It is possible to construct lines that are perpendicular to each of the circular sections and also go through the center of the indicatrix (Fig. 8-5). These lines are *optic axes*. It is because there are two such lines that materials falling into this optical category are called *biaxial*. The acute angle

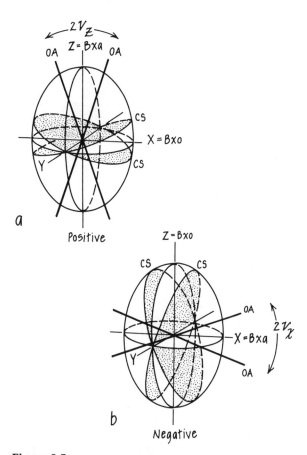

Figure 8-5.
A perpendicular line through the center of a circular section is called an optic axis. As biaxial materials have two circular sections, they also have two optic axes. (A) The acute angle between the two optic axes (called the optic angle or $2V$) is bisected by the Z vibration direction: Z is the acute bisectrix (Bxa) and X is the obtuse bisectrix (Bxo). This relationship defines a positive biaxial mineral. (B) The vibration direction X bisects the acute angle between the two optic axes. As X = Bxa and Z = Bxo, the mineral is defined as biaxial negative.

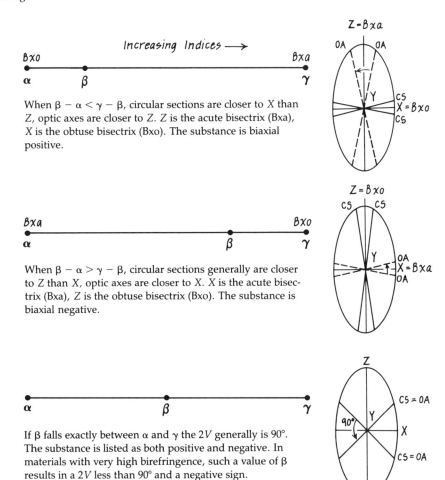

Figure 8-6.
Summary of the relations between the value of β relative to α and γ.

When $\beta - \alpha < \gamma - \beta$, circular sections are closer to X than Z, optic axes are closer to Z. Z is the acute bisectrix (Bxa), X is the obtuse bisectrix (Bxo). The substance is biaxial positive.

When $\beta - \alpha > \gamma - \beta$, circular sections generally are closer to Z than X, optic axes are closer to X. X is the acute bisectrix (Bxa), Z is the obtuse bisectrix (Bxo). The substance is biaxial negative.

If β falls exactly between α and γ the $2V$ generally is $90°$. The substance is listed as both positive and negative. In materials with very high birefringence, such a value of β results in a $2V$ less than $90°$ and a negative sign.

between the optic axes usually is called the *optic angle* or $2V$. In certain cases the optic angle may refer to the angle between the optic axes when measured about a specific vibration direction, for example $2V_z = 42°$; in these cases, the subscript X or Z is used and the $2V$ may exceed $90°$. The optic angle, which can either be measured or approximated when identifying unknown materials, is a very useful parameter in the determination of unknown biaxial materials.

The value of the index β relative to α and γ determines the *optic sign* (Fig. 8-6). When β is closer in value to α than it is to γ, the substance is termed *biaxial positive*. When β is closer in value to γ than it is to α, the substance is generally termed *biaxial negative*. If the value of β falls exactly between α and γ, the substance is listed in both positive and negative categories. The optic sign is more precisely defined below. Note that as β and α approach each other in value, the biaxial positive indicatrix approaches the shape of a uniaxial positive indicatrix. Similarly, as β approaches γ, the biaxial negative indicatrix approaches the shape of a uniaxial negative indicatrix.

Consider the consequences of the value of β relative to α and γ. If β is closer to α than to γ, this causes the circular sections to make a smaller angle to X than to Z (Fig. 8-4). The optic axes in turn have a smaller angle to Z than to X. In this case Z bisects the acute angle between the optic axes, and X bisects the obtuse angle. For such materials Z is called the *acute bisectrix* or *Bxa*, and X is called the *obtuse bisectrix* or *Bxo*. A biaxial positive mineral is precisely defined as such by stating that Bxa = Z and Bxo = X. The reverse relationship is true for a biaxial negative mineral (Bxa = X and Bxo = Z). When the $2V$ is 90°, these definitions no longer hold.

The optic angle can be calculated if the values of the three indices α, β, and γ are known, by using the following equation:

$$\cos V_z = \frac{\alpha}{\beta} \sqrt{\frac{(\gamma + \beta)(\gamma - \beta)}{(\gamma + \alpha)(\gamma - \alpha)}}$$

where V_z is the angle between one optic axis and the Z direction. For a positive mineral the $2V_z$ is less than 90°. For a negative mineral, where X is the acute bisectrix, $2V_z$ is greater than 90° and the supplement of the angle is taken to obtain the usual acute angle for $2V$. For minerals of low birefringence, when β is exactly between α and γ in value the $2V$ is almost 90°. However, in the case of a material of high birefringence this is not necessarily true; for example, given $\alpha = 1.50$ and $\gamma = 1.90$, a value of 1.70 for β results in a $2V$ of 80° and a negative sign, as can be calculated with the equation above.

A graphical method of determining $2V$ from the refractive indices was devised by J. B. Mertie (1942, p. 538). His nomogram is given in Figure 8-7 and its use demonstrated in Figure 8-8 (pp. 86 and 87). In addition, either the Mertie chart or the above equation can be used to determine an unknown index if the $2V$, sign, and the other two indices are known. Examination of Figures 8-7 and 8-8 shows clearly how the $2V$ varies as a result of the numerical relation of β to α and to γ. The $2V$ is 90° when the value of β is close to being halfway between the values of α and γ. As the value of β approaches either α or γ, $2V$ decreases toward 0°. If β coincides with either α or γ, the $2V$ becomes zero, and only two principal indices of refraction are present. We thus can regard a uniaxial substance as a special type of biaxial material having a $2V$ of 0°.

VIBRATION DIRECTIONS IN BIAXIAL CRYSTALS

The ray velocity surfaces for biaxial crystals are ellipsoidal. Consequently, incident light entering a biaxial crystal with either perpendicular or nonperpendicular incidence is usually bent slightly from the path suggested by Snell's law. This sets up a significant number of complications which will generally be ignored here, as their aggregate effect is minor for our purposes.[1] It should be kept in mind, therefore, that the following discussions, although providing good approximations, deviate slightly from the whole truth. This is of course the case for many discussions, whether or not a disclaimer is present.

We shall first consider the vibration directions and their indices of refraction for some different crystal orientations. These orientations will be the following: two principal planes vertical, one principal plane vertical, no principal plane vertical.

Two Principal Planes Perpendicular to the Stage

The line of intersection of two principal planes is parallel to one of the three principal vibration directions (Fig. 8-1). Thus when two principal planes are perpendicular to the microscope stage, one of the three principal optical directions (either X, Y, or Z) is also perpendicular to the stage. The other two principal optical directions are par-

---

1. This subject is discussed in optical mineralogy texts such as Bloss (1961), Hartshorn and Stuart (1969), and Wahlstrom (1979).

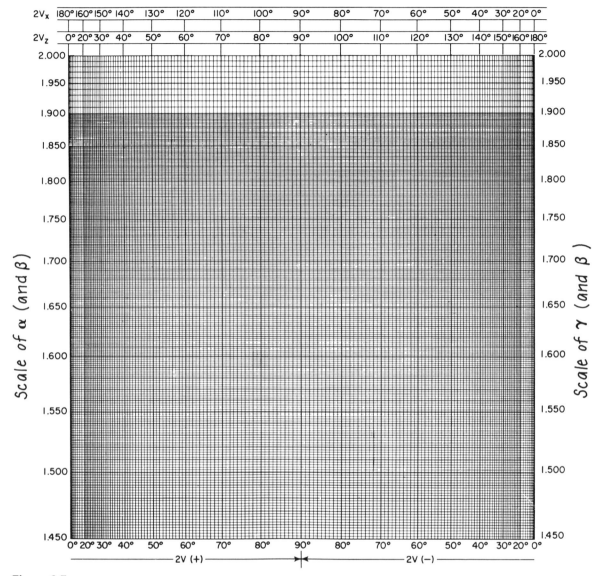

Figure 8-7.
Relationship between values of α, β, γ, 2V, and optic sign (Mertie, 1942). Use of the chart is shown in Figure 8-8.

Vibration Directions in Biaxial Crystals

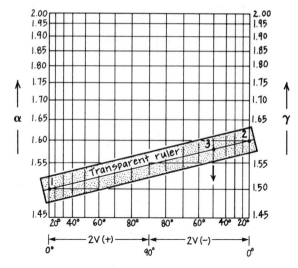

Figure 8-8.
Determination of the 2V and sign with the Mertie chart (Fig. 8-7). A transparent ruler is laid on the chart so as to correspond in position with α and γ (points 1 and 2). From the value of β (point 3), move downward to the base and read the sign and value of 2V on the horizontal scale.

allel to the microscope stage and are permissible vibration directions. Let us have Bxa be perpendicular to the stage and both Y and Bxo parallel to the stage (Fig. 8-9A). The vibration direction that corresponds to the Y direction yields the index β; the vibration direction parallel to Bxo yields an index of refraction that is either α (for positive minerals) or γ (for negative minerals). The situation is similar to that of a uniaxial crystal whose optic axis is parallel to the stage (Fig. 5-4B); here, two vibration directions with indices of ω and ε are parallel to the stage.

The two permissible vibration directions in Figure 8-9B are in diagonal orientation. As neither direction is parallel to the incident EW plane-polarized light, the light transmitted through the crystal is constrained to vibrate in the two diagonal vibrational planes. Retardation occurs and the crystal produces interference colors when viewed between crossed nicols. If the stage is rotated until either of the permissible directions becomes

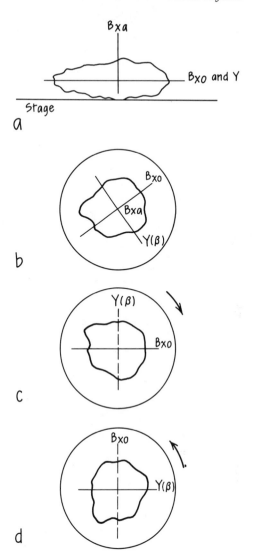

Figure 8-9.
(A) A grain oriented with Bxo and Y parallel to the stage and Bxa normal to the stage. (B) When viewed between crossed nicols, the grain in diagonal orientation shows interference colors. (C) The grain, when rotated clockwise, goes to extinction when Bxo is EW. With uncrossed nicols the Bxo index (either α or γ) can then be estimated with a Becke test. (D) The grain when rotated counterclockwise goes to extinction when Y is EW. After uncrossing the nicols, the β index can be estimated with a Becke test.

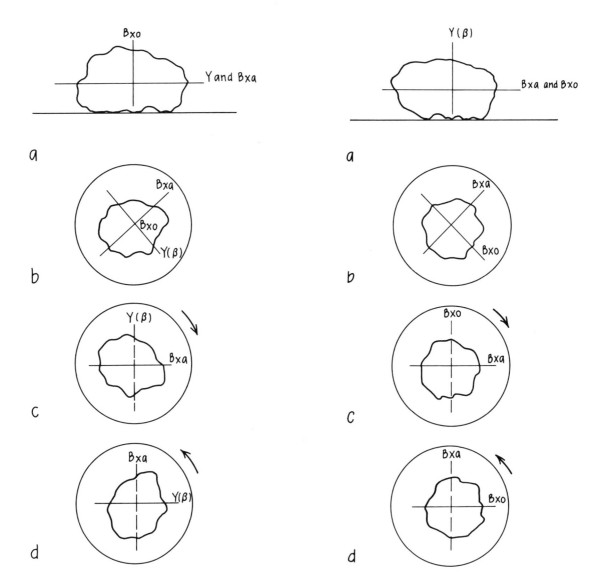

Figure 8-10.
(A) A grain oriented with Y and Bxa parallel to the stage and Bxo normal to the stage. (B) With crossed nicols, the grain in diagonal orientation shows interference colors. (C) The grain, when rotated clockwise, goes to extinction when Bxa is EW. With uncrossed nicols, the Bxa index (either α or γ) can be estimated with a Becke test. (D) The grain, when rotated counterclockwise, goes to extinction when Y is EW. After uncrossing the nicols, the β index can be estimated with a Becke test.

Figure 8-11.
(A) A grain oriented with Bxa and Bxo parallel to the stage and Y normal to the stage. (B) With crossed nicols, the grain in diagonal orientation shows interference colors. (C) The grain, when rotated clockwise, goes to extinction when Bxa is EW. With uncrossed nicols the Bxa index (either α or γ) can be estimated with a Becke test. (D) The grain, when rotated counterclockwise, goes into extinction when Bxo is EW. After uncrossing the nicols, the Bxo index (either γ or α) can be estimated with a Becke test.

parallel to the EW plane-polarized light (Fig. 8-9C and D), the light maintains its single EW vibration during transmission through the crystal. Such light is eliminated by the upper polarizer and the crystal is at extinction. Removal of the upper polarizer (after attainment of extinction) permits a check on the index of refraction, either the index corresponding to Bxo (part C), or the β index (part D).

The situation is similar when either the Bxo or Y direction is perpendicular to the stage (Figs. 8-10 and 8-11). Here again, when one of the principal optical directions is perpendicular to the stage, the other two are parallel to the stage. During stage rotation between crossed nicols the grain goes to extinction when one of the permissible vibration directions is parallel to the plane-polarized incident EW light. In such an orientation, a principal index can be estimated.

One Principal Plane Perpendicular to the Stage

When one principal plane is perpendicular to the stage, a large variety of crystal orientations is possible.

The upper line in Figure 8-12 shows an indicatrix in NW–SE vertical cross section on a horizontal microscope stage. The indicatrix is rotated clockwise about Y (from part A through E). Beside each indicatrix is a mineral grain as seen in the microscope, with its permissible vibration directions indicated in the diagonal directions. We begin with two principal planes vertical, as described above; the two permissible vibration directions in part A permit measurement of the α and β indices. In part B the grain has been rotated 30° about the Y direction so that an intermediate index α' is parallel to the stage; the measurable indices are now α' and β, and only the XZ plane is vertical. Additional rotation (part C) causes the circular section to become parallel to the stage. Here, the value of α' has increased until it is equivalent to that of β. In this case only a single index, β, can be measured on the crystal (as an optic axis is vertical). Additional rotation in part D brings a γ' radius parallel to the stage, thus permitting index measurements of γ' and β indices.

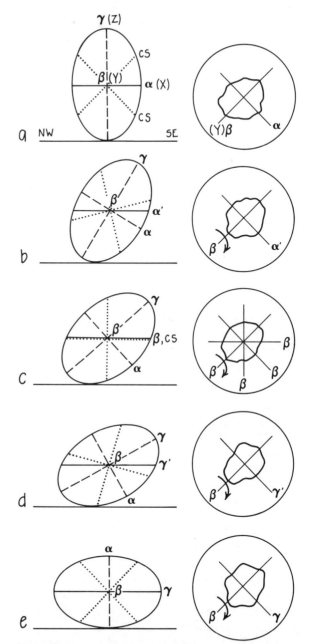

Figure 8-12.
Progressive inclination of a biaxial grain about the Y direction, with the XZ plane perpendicular to the stage. In the left diagrams the grain is represented by an indicatrix in vertical cross section, and is viewed from the SW. The lower right show the grain as viewed in the microscope. The effect of changing grain orientation is described in the text.

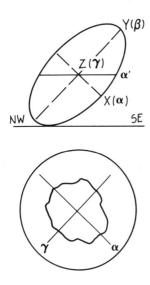

Figure 8-13.
A grain (represented as an indicatrix) oriented with one principal direction (here Z) parallel to the stage can be used to estimate the index associated with that direction (γ) and an intermediate index (α').

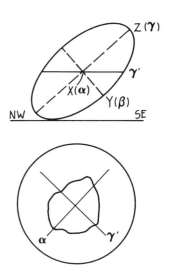

Figure 8-14.
A grain (represented as an indicatrix) oriented with one principal direction (here X) parallel to the stage can be used to estimate the index associated with that direction (α) and an intermediate index (γ').

Continued rotation about Y (to part E) brings a second (XY) principal plane vertical and permits measurement of γ and β. The above example shows that when a single principal plane is perpendicular to the microscope stage, one of the two permissible vibration directions will be a principal index direction (in this case β); the other permissible vibration direction will have an index of refraction between the two principal indices within the vertical principal plane (in this case between α and γ). The principal index direction that remains horizontal is perpendicular to the vertical principal plane.

The situation is similar when other single principal planes are perpendicular to the stage. In Figure 8-13, the XY principal plane is vertical and Z is parallel to the stage. The upper part of the figure shows the orientation of the indicatrix in a vertical NW–SE section. The permissible vibration directions are oriented so as to permit measurement of γ and α' (after appropriate stage rotation), as seen below in the field of view of the microscope.

The third situation, a vertical YZ plane perpendicular to the stage, is shown in Figure 8-14. Here the orientation of the grain permits measurement of α and γ' after appropriate stage rotation.

No Principal Plane Perpendicular to the Stage

This is the orientation that occurs most commonly, particularly for fragments that do not possess significant cleavages (Fig. 8-15). As none of the three principal vibration directions is parallel to the stage, it is possible only to make a Becke test of intermediate indices (α' and γ'). Although this is a less than perfect situation, such measurements are often sufficient to reduce the number of reasonable possibilities when identifying an unknown material.

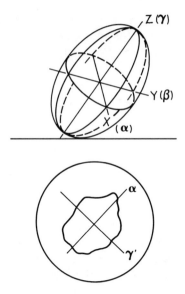

Figure 8-15.
A grain (represented as an indicatrix) oriented with none of the three principal directions parallel to the stage can be used to estimate α' and γ'.

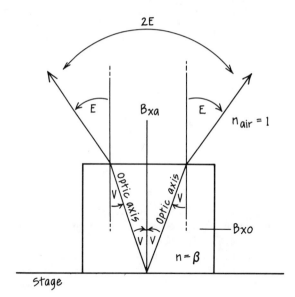

Figure 8-16.
A grain oriented with Bxa normal to the stage is viewed with conoscopic illumination. Some of the divergent light rays from below travel parallel to the two optic axes and therefore have an index of β. Upon leaving the grain these rays are refracted at the air-grain interface. The angle between the two rays within the grain is $2V$; after refraction at the top of the crystal, the angle between the two rays is $2E$.

BIAXIAL INTERFERENCE FIGURES

The previous discussion, although describing relationships between various crystal orientations and index measurements, does not provide a method of distinguishing among the various orientations. Orientation must be determined with biaxial interference figures, obtained in the same way as those of uniaxial crystals (Chapter 6).

When two principal planes are perpendicular to the stage, either the Bxa, the Bxo, or Y may be vertical. Corresponding interference patterns are called either Bxa, Bxo, or optic normal (O.N.) interference figures. The optic normal figure is so called because the Y direction is normal to the *optic axial plane* (the XZ plane in the indicatrix, which contains the two optic axes).

Acute Bisectrix (Bxa) Interference Figure

When Bxa is perpendicular to the stage and Bxo and Y are parallel to the stage, the two optic axes are in a vertical (XZ) plane. The question immediately arises as to whether or not the melatopes (which represent the optic axes) will be within the field of view; the answer to this is critical, as knowledge of the location of the two melatopes permits a determination of the value of $2V$. It is necessary to digress and consider possible locations of the melatopes.

Figure 8-16 shows a vertical cross-section of the XZ plane of a mineral on a horizontal microscope stage. The illumination is conoscopic. Some of the convergent incident polarized light travels along the optic axes. This light has the index β (as the

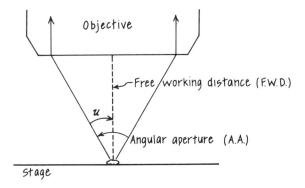

Figure 8-17.
An objective lens is focused on a grain on the microscope stage. The distance between the objective and the object in focus is called the free working distance (F.W.D.) of the lens. The maximum angle of divergent rays that can enter the objective lens at its free working distance is called the angular aperture, A.A. (half of which is called u).

vibration directions are in the circular sections, which are perpendicular to the optic axes). Before this light reaches the objective lens it must pass through air. Refraction occurs at the grain-air interface. The light will enter the air at an angle of refraction that is called E, rather than V. This relationship is described by Snell's law:

$$n_1 \sin \sphericalangle_1 = n_2 \sin \sphericalangle_2$$

The index n_1 is the β index within the crystal. The incident angle $\sphericalangle_1$ within the crystal at the crystal-air interface is equal to V. The index n_2 is the index within the second medium (here air), and $\sphericalangle_2$ is the refractive angle in medium 2 (here E). Therefore

$$\beta \sin V = (1) \sin E$$
$$\sin E = \beta \sin V$$

Knowing the index β and V (or $2V$), the angle E (or $2E$) can be calculated. This is the actual angle "seen" by the microscope, rather than the $2V$ value listed in the determinative tables.

We must now determine the maximum angular spread of light rays that can be transmitted by the objective lens. This angle is a function of the numerical aperture (N.A.) of the objective lens, where

$$\text{N.A.} = n \sin u = n \sin (\text{A.A.}/2)$$

Where A.A. is the angular aperture, u is A.A./2, and n is the lowest index of refraction between the object and the objective.

What does all this mean? Consider the cross-section of Figure 8-17, showing an objective lens and a grain on the microscope stage. When the grain is in focus, the objective is at a fixed vertical distance from it. This distance is the focal distance or free working distance (F.W.D.) of the lens. Higher power objectives have a smaller F.W.D. than lower power objectives. Rays of light from the grain can enter the lens only within an angular limit that is dependent upon the F.W.D. and the cross-sectional diameter of the lens. The angular spread of rays that the lens can admit is called the angular aperture (A.A.); half the angular aperture is called u. Using the equation above and the value of the numerical aperture, which is generally printed on the side of the lens, it is possible to calculate u and in turn the A.A. For the usual lenses, n is taken as 1, the index of refraction of air.[2]

A high-power objective with numerical aperture of 0.85 is rather common. Using the equation above, it can be calculated that the angular diameter of the field of view (A.A.) is 116°:

$$\text{N.A.} = n \sin u$$
$$0.85 = (1) \sin u$$
$$\sin u = 0.85$$
$$u = 58°$$
$$\text{A.A.} = 116°$$

---

2. For higher than average magnifications it is possible to use oil-immersion objectives. Here a drop of liquid is placed in contact with the objective lens and the top of the cover slip. The value of n in the equation is that of the liquid.

Therefore, if a mineral yields a Bxa interference figure with a 2E of 116°, the two melatopes would be located at the edge of the field of view. A larger 2E would put the melatopes outside of the field, whereas a smaller 2E would put them within the field.

Figure 8-18 graphs the relationship between 2V, β, and the numerical aperture of two widely used high-power objectives. If the high-power objective in your microscope has a numerical aperture of 0.85, the angular aperture is 116°. Assume that a Bxa interference figure has melatopes that lie at the edge of the field (2E = 116°). If the index β of the mineral has been determined to be 1.50, the graph shows that the 2V is 69°. With the same objective and a mineral with β index of 1.65 and 2E = 116°, the 2V would be 62°. If the numerical aperture of your high-power objective is not one of those shown in Figure 8-18, it would be very useful to calculate its angular aperture and the values of 2V and β that locate the melatopes at the edge of the field of view (using the equations given above).

The preceding discussion is of more than academic (worthless) interest, because correct estimation of 2V is of great significance in identifying biaxial minerals. Consider Figure 8-19, which shows the location of the melatopes of the same mineral when viewed with objectives with numerical apertures of 0.65 and 0.85; the apparent angular difference between the pairs of melatopes is significant. The numerical aperture of the objective lens cannot be ignored.

Returning again to the Bxa figure, assume that the 2E of a mineral is small enough that both melatopes fall within the field of view, as shown in plan view and vertical cross section in Figure 8-20A. Part B of the figure shows the principal vibration directions and melatopes (optic axes) in an incomplete interference figure. Notice that Bxa, Bxo, and the two melatopes all fall on the same vertical (NE–SW) plane. The task now is to deduce the orientation of the permissible vibration directions within the field of view, as well as their respective amounts of retardation.

The privileged (permissible) vibration direc-

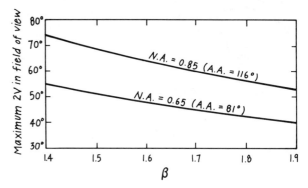

Figure 8-18.
A chart showing the maximum 2V that can be observed in a Bxa interference figure with various combinations of β and numerical aperture (N.A.). For example, a Bxa interference figure is obtained from a substance whose β index is 1.6. If the melatopes appear at the edge of the field of view when observed with an objective whose numerical aperture is 0.85, the 2V is about 64°; if the N.A. is 0.65, the 2V is about 48°. A larger 2V would result in both melatopes being outside of the field of view, and a smaller 2V would place the melatopes within the field of view.

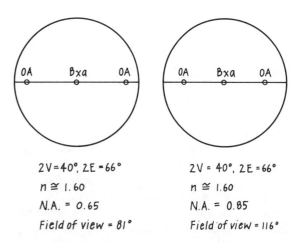

Figure 8-19.
The position of the melatopes in a Bxa figure changes as a function of the numerical aperture of the objective that is used.

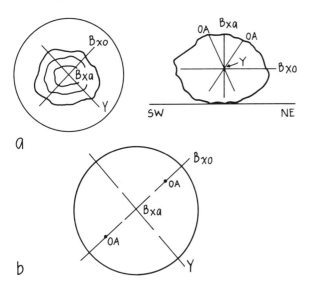

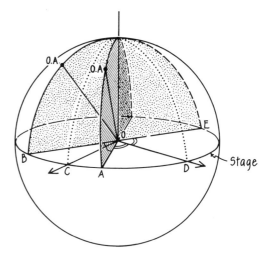

Figure 8-20.
Finding the general location of the melatopes in a Bxa interference figure. (A) A grain is oriented with Bxa normal to the stage; the Bxo vibration direction is oriented NE–SW. In vertical cross section it can be seen that the optic axes fall within the Bxa–Bxo plane, as is true for all biaxial materials (see Fig. 8-5). (B) In the corresponding Bxa interference figure, the melatopes lie within the NE–SW plane, their particular location dependent upon the values of β, $2V$, and N.A.

Figure 8-21.
Biot–Fresnel technique for finding permissible vibration directions in interference figures at the center of the field of view. The equatorial plane is taken as the microscope stage, and the microscopist's line of sight is along the vertical axis. The optic axes of a biaxial material (not shown) lying on the stage at O are along the two lines between O and O.A. Erect two vertical planes (darkened) that include the vertical line of sight and the two optic axes. Their traces on the microscope stage are indicated as lines OA and OB. The permissible vibration directions for a vertical ray are found by bisecting the trace of the two vertical planes on the microscope stage. One direction is OC, bisecting the angle BOA, and the other is OD, bisecting angle AOE.

tions at any point on the interference figure can be determined by an approximation of the *Biot–Fresnel rule*, which states that the vibration directions within a crystal are the bisectors of the angles of intersection made by two imaginary planes, each of which contains an optic axis and the observer's line of sight. This is not as bad as it sounds. The relationship can be visualized with the aid of Figure 8-21. The microscope stage is shown as the equator of a sphere. The line from the top of the sphere to its center is the observer's line of sight, down the microscope tube. Two optic axes (labeled O.A.) intersect the surface of the sphere, and the two darkened planes each contain the observer's line of sight and a single optic axis. These planes intersect the stage along the lines OA and OB. OC, which bisects the acute angle between OA and OB, is one of the two permissible vibration directions for a crystal in this orientation. The second permissible vibration direction is OD, which bisects the obtuse angle between the same two vertical planes, whose horizontal traces are shown as OA and OE; the two permissible vibration directions OC and OD are, of course, mutually perpendicular.

The Biot–Fresnel rule permits us to determine the permissible vibration directions at any point on a biaxial interference figure. As an example, determine the vibration directions at a point such as A in Figure 8-22. The trick is to temporarily consider point A as being the line of sight (recalling that only a ray in the center of the field of view is perpendicular to the stage). Construct two lines, joining point A with each of the optic axes. Bisect the angles of intersection of these two lines (the two heavy lines through point A); these bisectors are the permissible vibration directions at A.[3] Figure 8-22B shows the same type of construction for points B and C. The construction lines drawn from each of the optic axes to the points of interest represent edge views of planes similar to those shown (darkened) in Figure 8-21.

If we carry out this analysis for many points, it is possible to derive a pattern of permissible vibration directions for the entire field of view (Fig. 8-23A). Lines can be drawn that connect the permissible vibration directions; the resultant pattern (Fig. 8-23B) is called a *skiodrome*. The only exceptions to the skiodrome pattern occur at the optic axes. Light traveling parallel to the optic axes behaves as if the crystal were isotropic, just as with light traveling parallel to the *c* axes in uniaxial grains. Consequently, light at the optic axes is plane-polarized EW, in correspondence with the incident light from below.

Note that skiodromes show the vibrations that are permitted by the grain, not necessarily those that actually exist. The vibrations that exist depend upon the direction of polarization of the incident light.

We can see an example of this by returning to Figure 8-22B. The incident light from the lower polarizer is plane-polarized in the EW plane. As neither permissible vibration direction (heavy lines) at point B corresponds to the EW vibration direction, the incident light is resolved into two vibration directions. This doubly polarized light undergoes retardation and passes through the upper polarizer, to show up as a particular color.

---

3. To do the construction accurately, all points should be plotted on a stereographic projection. The dashed lines drawn in Figure 8-21 are actually great circles. The approach here, however, yields a fair approximation of the permissible vibration directions.

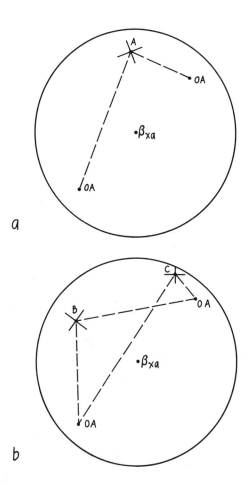

Figure 8-22.
Approximation of the Biot–Fresnel technique for finding the permissible vibration directions at any point on an interference figure. The positions of the melatopes (labeled O.A.) are indicated. (A) In order to find the vibration directions at point A, draw lines (dashed) from each of the melatopes through point A. Bisecting the angles of intersection of the dashed lines at A yields the two permissible vibration directions (heavy lines). (B) Determination of the permissible vibration directions at points B and C. This approximation uses straight lines rather than great circles on a stereographic projection.

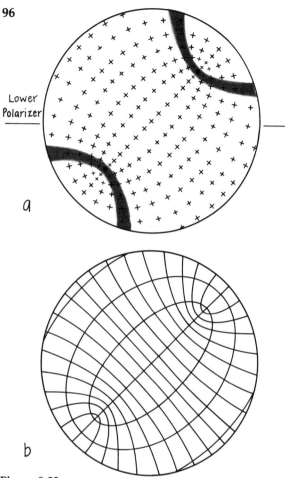
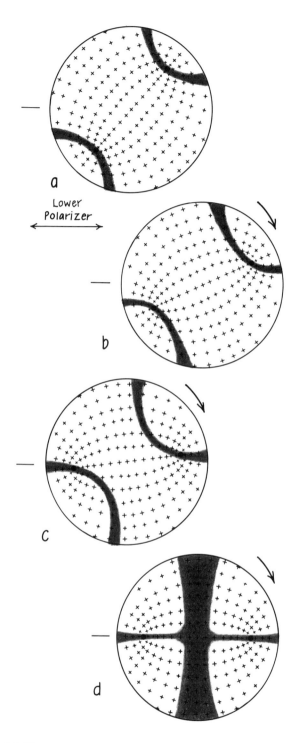

Figure 8-23.
(A) Composite sketch of the permissible vibration directions within a Bxa figure. In some parts of the field (darkened), one permissible vibration direction is parallel to the privileged vibration direction of the lower polarizer. In these regions retardation does not occur; instead, isogyres are formed. (B) Skiodrome of a Bxa figure. Permissible vibration directions at any point are parallel to the two sets of lines.

Figure 8-24.
Isogyre movement in a Bxa figure during stage rotation. (A) In the diagonal position, the isogyres have a maximum separation and are symmetrically disposed to the optic plane (containing the two optic axes). (B) Clockwise rotation of the stage results in counterclockwise rotation of the isogyres about the optic axes. The isogyres are not symmetrically disposed to the optic plane. (C) With continued clockwise rotation, the isogyres approach each other. (D) The isogyres join and form a cross when the optic plane is EW. The isogyre cross forms four times during a 360° rotation of the stage.

At point C, on the other hand, one permissible vibration direction corresponds to that of the incident light. Therefore, the incident light is transmitted through the crystal plane-polarized EW, then is eliminated by the upper polarizer (with its NS privileged direction). The point C is part of an isogyre.

Using the described approach it becomes possible to determine the position of isogyres and their motion during rotation of the stage. In Figure 8-23A the dark arcs represent regions where one of the two privileged directions is EW. These areas are converted into isogyres by the action of the upper polarizer.

Rotation of the stage causes rotation of the permissible vibration directions, thus the isogyres shift in position (Fig. 8-24). The isogyres pivot around the two optic axes in a direction opposite that of the stage rotation. When the optic axes are located on EW or NS vertical planes, the two isogyres intersect to form a cross (Fig. 8-24D). The isogyres never leave the optic axes, because only plane-polarized EW vibrations exist at these locations. Note that the isogyre that joins the two optic axes is considerably thinner than the other isogyre that is normal to it; this contrasts with the two equally thick isogyres of a uniaxial optic axis figure. (See Figure 5 of Plate 2.)

The separation of the two isogyres is greatest when the vertical plane containing the two optic axes is brought into a diagonal position (Fig. 8-24A). The extent of this maximum separation is determined by the $2V$ and β index of the mineral and the angular aperture of the objective lens. With a large $2V$, the optic axes may lie outside of the field of view of the microscope; in this case the isogyres leave the field during rotation. With a moderate to small $2V$, the isogyres remain in the field during rotation (Fig. 8-25).

Figure 8-25.
Isogyre positions for diagonally oriented Bxa figures. The grain has a β index of 1.60. The left column applies to an objective lens with a numerical aperture of 0.65, the right column to one with a numerical aperture of 0.85.

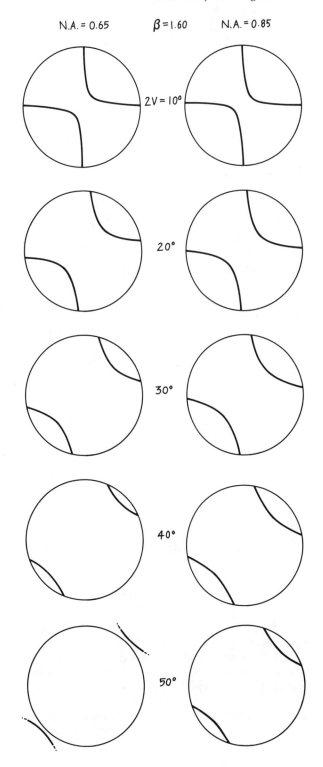

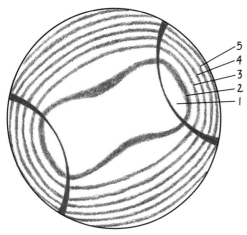

Figure 8-26.
A Bxa interference figure of moderate $2V$ in diagonal position. Interference colors increase from black at the melatopes to higher order colors at the edge of the field of view ($5 > 4 > 3 > 2 > 1$).

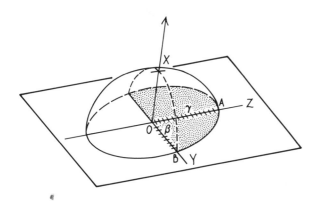

Figure 8-27.
A light ray travels parallel to the X direction of a grain in diagonal orientation on the microscope stage. The grain is represented as half of a biaxial indicatrix. Vibration directions (hatched lines in the darkened section) have the indices γ and β, and correspond to the maximum and minimum radii of the darkened section.

The particular *interference color* seen at any point within an interference figure is a result of the amount of retardation at that point. (The topic was introduced in Chapter 5.) Figure 8-26 and Plate 2 (Figs. 3, 4, and 5) show a possible distribution of interference colors in a Bxa figure. At the melatopes, which mark the positions of the optic axes, no color is present. With increasing angular distance from each melatope the interference colors rise to form a series of concentric isochromatic curves, similar to those in uniaxial figures. The two sets of concentric curves blend together between the melatopes and toward the perimeter of the field of view.

A partial explanation for the interference color phenomenon is illustrated in Figures 8-27 through 8-30. Each figure shows the upper surface of a negative biaxial indicatrix (Bxa vertical) on a plane that represents the microscope stage. In Figure 8-27 a ray travels vertically through the crystal. The permissible vibration directions (at right angles to the ray's direction of travel) are shown as the maximum and minimum radii OA and OB; the relative lengths of these lines represent the indices of refraction of the two vibrations γ and β. The retardation developed is $t(\gamma - \beta)$, with t the grain thickness.

Figure 8-28 shows a ray traveling at a small angle from the vertical within the XZ plane. The maximum and mininum radii within the shaded plane (approximately perpendicular to the ray) are the two permissible vibration directions OA' and OB. The lengths of these radii represent the indices of refraction of these vibrations, γ' and β. By changing the ray direction within the XZ plane from the stage normal, the index of refraction of the vibration within that plane has changed from γ to γ'. The retardation has decreased to $t(\gamma' - \beta)$, with a resultant decrease in interference color. Thickness (t) is regarded as a constant.

Continued change of the ray direction within the XZ plane brings the ray direction into coincidence with the optic axis (Fig. 8-29). The associated vibration directions are now within the circular section (which is perpendicular to the optic

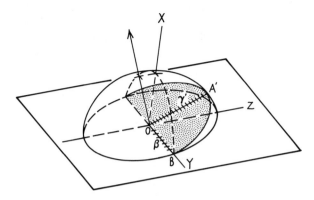

Figure 8-28.
A light ray travels in a direction between X and Z within a grain (represented as half of a biaxial indicatrix) in diagonal orientation. Vibration directions correspond to the maximum and minimum radii of the darkened section, and have the indices γ' and β. The ray path is not exactly perpendicular to the γ' vibration direction.

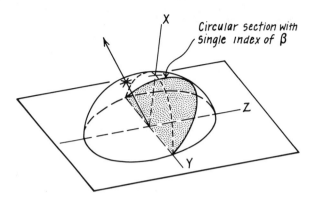

Figure 8-29.
An inclined light ray travels along an optic axis within a grain (represented as half of a biaxial indicatrix) in diagonal orientation. The vibrations are within the circular section, the radii of which have the index β.

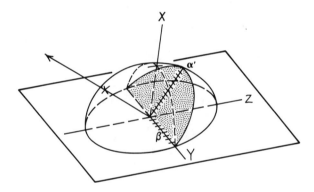

Figure 8-30.
An inclined light ray travels at a large angle to X within the XZ plane. Vibration directions within the darkened plane correspond to the maximum and minimum radii β and α'.

axis). As any vibration within the circular section has a fixed index of β, no retardation occurs, so no interference colors are developed. Such a ray marks the position of a melatope on the interference figure.

A still greater inclination of the ray (Fig. 8-30) brings the vibration direction within the XZ plane between the circular section and X. This vibration has the index α', and the retardation has increased to $t(\beta - \alpha')$. Further inclination of the ray would result in greater retardation until the value of α' decreases to α.

After this it may pay to re-examine Figure 8-26, which shows a typical color distribution within a Bxa figure. Consider the NE–SW diagonal direction; intermediate colors are present in the center of the field, decrease to a minimum at the melatope, and then rise again toward the edge of the field of view, corresponding to the conclusions drawn from Figures 8-27 through 8-30. Interference colors in other parts of the field of view may be derived in a similar manner. The particular level of colors seen in a Bxa figure will vary as a function of the birefringence and thickness of the grain being examined. Increase of either birefringence or thickness, or both, results in higher level colors. The $2V$ determines the particular placement of the colors.

Obtuse Bisectrix (Bxo) Interference Figure

The Bxo figure resembles a Bxa figure with a large $2V$. In fact, the two figures become identical when the $2V$ is 90°.

Consider first the location of the isogyres. The permissible vibration directions can be determined, as for the Bxa figure, by means of the Biot–Fresnel rule and the construction of a skiodrome. One is shown in Figure 8-31. When one of the two permissible vibration directions is parallel to the incident EW vibrations, only a single EW vibration is present at this point, and an isogyre is created by the upper polarizer. Where neither permissible direction is parallel to the incident EW vibrations, light is resolved into two components with different velocities, retardation occurs, and interference colors form. As the microscope stage is rotated, the pattern of permissible vibration directions is also rotated, causing shifts in the positions of the isogyres.

The movements of the isogyres follow a pattern like those of a Bxa figure with a large $2V$. Rotation brings the two isogyres into the field along diagonal paths, to form a cross with EW- and NS-trending arms (Fig. 8-32). Further rotation causes

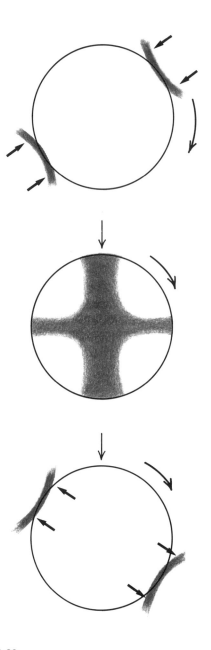

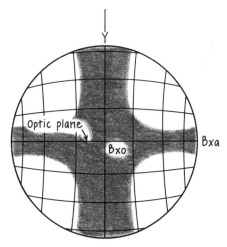

Figure 8-31.
Skiodrome of a Bxo figure. The darkened area corresponds to the position of the isogyre cross.

Figure 8-32.
Determining grain orientation in a Bxo figure. With rotation of the stage, isogyres enter in opposite quadrants, form a cross, then leave in the remaining quadrants. The Bxa direction is in the two quadrants within which the isogyres leave the field. Y is at 90° to Bxa.

the cross to break into two isogyres that leave the field along the opposite diagonals. The isogyres always leave the field during stage rotation. Interference colors are lowest in the quadrants within which the isogyres have left the field.

Optic Normal (O.N.) Figure

The biaxial optic normal or flash figure is identical in appearance and utility to uniaxial optic normal figures (Chapter 6). Note that during stage rotation, the isogyres leave the field of view in the quadrants that contain the Bxa.

Interference Colors and Figure Type

It is often desirable or necessary to obtain a particular type of interference figure. This can be facilitated by noticing the level of interference colors displayed by a variety of grains of roughly equal thickness.

The maximum interference colors are obtained from grains in which the principal optical directions X and Z are parallel to the stage. Such grains yield an optic normal (O.N.) interference figure. Reference to Figure 8-6 will show that Bxo figures are produced by grains with higher interference colors than those grains that produce Bxa figures; this is because the difference between β and the index that represents Bxa is always larger than the difference between β and the index that represents Bxo. Grains that yield an optic axis (O.A.) figure have a single vibration direction and are in constant extinction. The relationships are summarized in Table 8-1. In general, the most information is obtained by examination of Bxa and O.A. interference figures, as they not only allow determination of sign, but also an estimate of $2V$ and type of dispersion.

Distinguishing among Centered Bxa, Bxo, and O.N. Figures

An acute bisectrix figure with moderate to small $2V$ presents no problem in recognition: during

Table 8-1
Interference figures and interference colors of grains

Level of interference colors observed	Type of interference figure obtained	Vibration directions parallel to the stage
Maximum	O.N. (flash)	Bxa, Bxo
Relatively high	Bxo	Bxa, O.N. (β)
Relatively low	Bxa	Bxo, O.N. (β)
No interference colors	O.A.	O.N. (β)

rotation of the stage, the isogyres remain constantly in the field of view or leave it only briefly. If, however, the isogyres leave the field fairly rapidly and remain out of view for a considerable amount of rotation, it becomes necessary to decide whether the figure is a Bxa (with a large $2V$), a Bxo, or an optic normal type.

Fortunately the problem has been solved (Kamb, 1958). In order to obtain the correct answer, it is necessary to know the numerical aperture of the objective being used, the average index of refraction of the mineral, and the number of degrees of stage rotation within which the isogyres remain in the field of view. The numerical aperture is stamped on the side of the objective. The average index of the mineral is $(\alpha + \gamma)/2$; this can be approximated and still yield reasonable results. Measure the number of degrees of stage rotation from the point at which the isogyres enter at the edge of the field, through the point where they form a cross, to the point where they leave the field (Fig. 8-32). This rotational angle, R, is used with Figure 8-33 to determine the type of interference figure, and simultaneously to obtain an approximate value of $2V$.

If the interference figure is slightly off-center, the two isogyres may enter and leave the field at slightly different times (Fig. 8-34). To obtain angle R, simply note the stage degree readings when each isogyre enters the field, then average the two readings to obtain the average point of entrance.

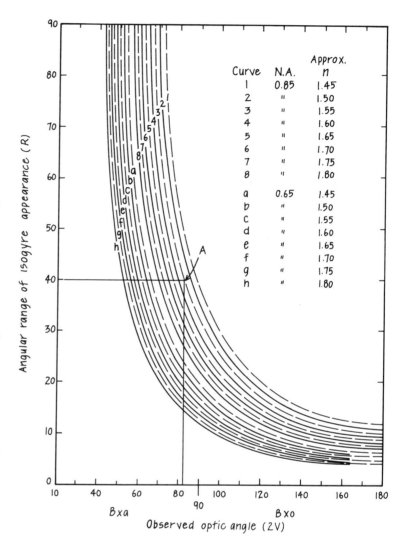

Figure 8-33.
Chart that permits determination of both the type of interference figure and the 2V (Kamb, 1958). Obtain a centered interference figure in which the isogyres enter and leave the field during stage rotation. Find the R value, the number of degrees of stage rotation in which the isogyres remain in the field of view. Proceed to the right from the R value (here 40°) to the curve that corresponds to both the numerical aperture of the objective and the average index of the grain (here point A on curve 3). From this point proceed downward and read the 2V at the base of the diagram. If the 2V is 90° or less, the interference figure is the Bxa type; if the 2V exceeds 90°, the figure is the Bxo type. If the value of R falls below the proper curve, the figure is of the optic normal type. *Be warned* that the R values obtained are often very questionable, and the chart should only be used to estimate approximate values of 2V.

Do the same with the two readings when the two isogyres leave the field. The difference between these two averages yields the angle R. Such averaging lends some uncertainty to the value of R, but generally produces correct results if the interference figure is close to being centered.

Armed with the rotational angle R, the numerical aperture of the objective, and the average index of the mineral under observation, turn now to Figure 8-33. Let us assume that the microscope has a numerical aperture of 0.85 and that the average index of a particular mineral is near 1.55. The curve to be used in the diagram is number 3. Assume that the rotational angle R is 40°. Find 40° on the vertical coordinate, then find where the curve crosses that level (at point A). The value on the horizontal coordinate directly below point A is 82°, which is the value of 2V.

The horizontal coordinate, the observed optic angle, ranges from 0° to 180°. If the determined value falls between 0° and 90°, the vertical principal index direction is Bxa; if the determined value is greater than 90°, the vertical index direction is Bxo. If the measurements are precise, the method permits a determination of 2V for either Bxa or Bxo figures. Values less than 90° are taken directly; values greater than 90° are subtracted from 180° to obtain the 2V. In practice, however, because the isogyres are often diffuse and the R measurement fairly inaccurate, the determined 2V should be considered a rough approximation. The precision can be improved by using a monochromatic light source.

The diagram also indicates whether Y is perpendicular to the stage. If the R value falls below the appropriate curve, the interference figure is the optic normal type. Determination of angle R is a very difficult measurement to make with an optic normal figure due to diffuseness of the isogyres, but it often is worth a try. As the 2V approaches 90°, the optic normal figure becomes increasingly diffuse and the isogyres remain in the field for smaller values of R. With a 2V of 90°, the isogyres enter and leave almost instantaneously during stage rotation.

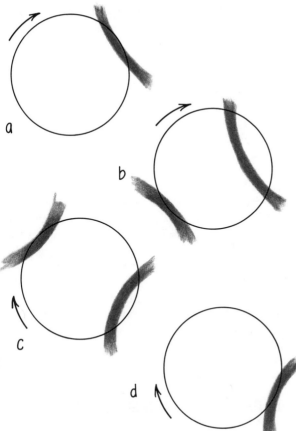

Figure 8-34.
Determination of angle R (Fig. 8-33) for slightly off-center interference figures. (A) Read stage vernier when the first isogyre enters the field (here 115°). (B) Read the vernier when the second isogyre enters the field (here 125°). Average the two readings to obtain an average "entering" value (here 120°). (C) Rotate through the isogyre cross position and read the stage vernier when the first isogyre leaves the field (here 140°). (D) Read the vernier when the second isogyre leaves the field (here 146°). Average these two values to obtain an average "leaving" value (here 143°). Angle R is the difference between the two averages (here 23°).

ESTIMATION OF 2V

Two methods of 2V estimation have been discussed. The first (Fig. 8-25) permits a general estimate based upon whether or not both melatopes are in the field of view when viewing a centered Bxa figure. The second, using the diagram of Kamb (Fig. 8-33), is based upon the range of stage rotation over which the isogyres remain in view.

In addition, as noted earlier, the amount of curvature shown by an isogyre at a melatope (in the diagonal position) is a function of the 2V. If the 2V is 90°, the isogyre passes through the melatope as a straight line. As the 2V approaches zero, the isogyre passes through the melatope with an almost 90° bend. Intermediate values of 2V yield intermediate curvatures. This relationship is summarized in Figure 8-35, which shows a variety of isogyres in interference figures in

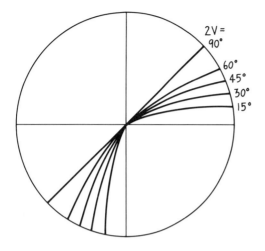

Figure 8-35.
Isogyre curvature of an optic axis interference figure as a function of $2V$.

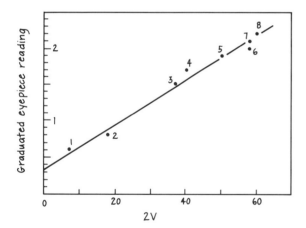

Figure 8-36.
A graduated eyepiece permits measured intermelatope distances in centered Bxa figures to be related to known values of $2V$. The diagram, from Parslow (1977), uses the following minerals: (1) phlogopite, (2) aragonite, (3) barite, (4) muscovite, (5) celestite, (6) gypsum, (7) diopside, (8) topaz. The points do not fall exactly in a straight line due to different values of β among the minerals.

which one of the two optic axes is perpendicular to the stage. (Note that when the $2V$ is small, a second isogyre may be within the field of view.) This method is useful for estimating $2V$ when an optic axis is near the center of the field.

Another approximate measurement of $2V$, discussed by Parslow (1977), uses a linear graduated eyepiece, one that contains a scale rather than the usual cross hairs. Also necessary are a group of thin sections with minerals covering a range of $2V$, each having a grain whose Bxa is perpendicular to the slide surface. These are often advertised by companies that make thin sections. The angular distance between melatopes in the 45° position is related to linear scale readings in the eyepiece. When these are plotted against the known values of $2V$ of the minerals (Figure 8-36), the more or less linear relation can be used to estimate the $2V$ of unknown minerals. The standard error of 2.4° is accurate enough for most mineral identifications. More precise methods are given in advanced texts such as Bloss (1961) and Wahlstrom (1979).

OFF-CENTER INTERFERENCE FIGURES

It is now necessary to reveal an unfortunate truth—some of the biaxial interference figures that you obtain will probably not be of the types described above. Rather than having two of the principal vibration directions parallel to the stage, it is common to have either a single direction or none at all. This is not all bad, as much information can be obtained from these figures if necessary.

Off-center figures can be considered by starting with a centered Bxa figure oriented in the diagonal position (Fig. 8-37A). Imagine that the crystal is rotated about a horizontal axis parallel to the Y direction; this causes the whole vibrational pattern and interference figure produced by the crystal to rotate as well. Several possible positions are shown in Figure 8-37B, C, and D. Consider these interference figures from the standpoint of symmetry.

The centered Bxa figure in Figure 8-37A has two diagonally oriented symmetry planes that pass through the center of the field of view. In parts B, C, and D, rotation of the crystal about the Y direction eliminates one of these symmetry planes: the figure has changed from being bisymmetric to monosymmetric. Notice that in parts B, C, and D, in spite of changing orientations, the axis of rotation, Y, is still parallel to the stage. This means that the β index can be measured with a Becke test when the stage is rotated into the proper position (45° counterclockwise). This horizontal principal index direction is perpendicular to the single plane of symmetry in the interference figures. In part A, where the interference figure is bisymmetric, horizontal principal index directions are perpendicular to both symmetry planes. Here both β and Bxo can be measured with a Becke test.

From this approach comes a method of analysis of interference figures. After obtaining an interference figure, rotate the stage until the figure is in a diagonal orientation. Observe the position of any diagonal symmetry planes that pass through the cross-hair intersection. Perpendicular to any symmetry plane in an interference figure there exists a principal index direction that is parallel to

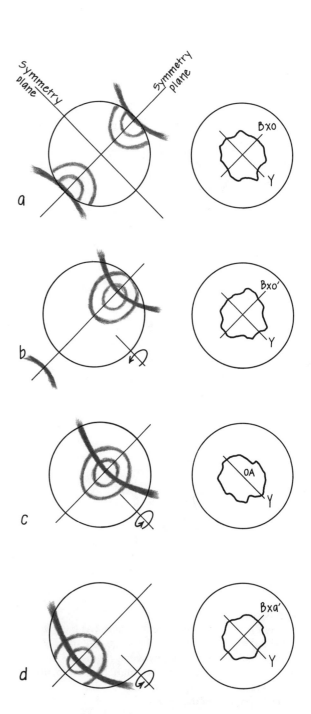

Figure 8-37.
Changes in grain and figure orientation with rotation about a principal vibration direction that is parallel to the stage (here Y). (A) Both Bxo and Y are parallel to the stage. The Bxa figure (in diagonal position) has two symmetry planes and can be referred to as a bisymmetric Bxa figure. (B) Rotation about Y eliminates a symmetry plane in the figure, which is now a monosymmetric Bxa figure. (C) Continued rotation about Y brings a melatope into vertical orientation. As is true with all optic axis figures, the symmetry is monosymmetric and a circular section is now parallel to the stage. (D) Continued rotation about Y inclines the melatope from the vertical, creating a monosymmetric Bxo figure. In all examples, perpendicular to each symmetry plane is a principal direction that is oriented parallel to the stage.

the stage, and therefore allows a measurement of either α, β, or γ (rather than of α' or γ').

Figure 8-38A shows a bisymmetric Bxa figure. Parts B and C show the effect of rotating the crystal around Bxo, which is parallel to the stage. The NW–SE symmetry plane is retained and the figure is monosymmetric. Using the rule given above, a single principal index direction is now known to be parallel to the stage; this index corresponds to Bxo (see Fig. 8-20B).

Least informative is a nonsymmetric Bxa figure (Fig. 8-39). Neither Bxo nor Y is horizontal, and no symmetry planes pass through the center of the field. Although not useful for precise index measurements, such figures can be used for a sign determination and estimation of $2V$.

Although Bxo and optic normal (O.N.) figures can be classified also as bisymmetric, monosymmetric, and nonsymmetric, they are more difficult to characterize than Bxa figures, because the isogyres are not in the field of view when the figure is in the diagonal position. It is necessary to note how the isogyres leave the field during rotation of the stage. If the figure is a bisymmetric Bxo type, stage rotation will cause the two isogyres to leave the field simultaneously; the last portions of the retreating isogyres are seen along a diagonal direction, as in Figure 8-40A. It follows that if the isogyres leave the field simultaneously they will both migrate equal distances from the

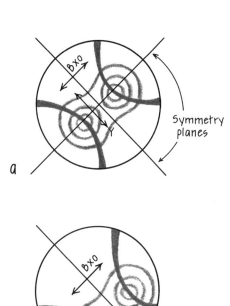

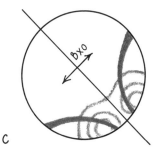

Figure 8-38.
A bisymmetric Bxa interference figure (A) has two principal indices parallel to the stage. Rotation of the grain about Bxo (B and C) converts the figure into a monosymmetric type. The orientation of the plane of symmetry indicates that the Bxo vibration direction remains parallel to the stage.

Figure 8-39.
Nonsymmetric interference figure shows that none of the three principal vibration directions is parallel to the stage.

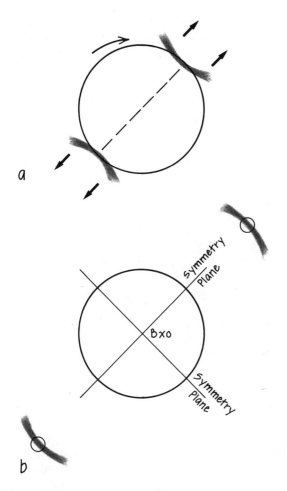

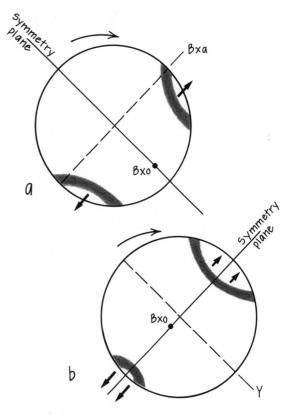

Figure 8-40.
A Bxo interference figure can be determined to be bisymmetric by noting how the isogyres leave the field of view during stage rotation. (A) Both isogyres leave the field simultaneously. The last bit of darkness is seen along a diagonal line (dashed) that joins opposite quadrants. (B) It can be assumed that the isogyres are in symmetrical arrangement when the figure is in diagonal orientation. Two planes of symmetry are present.

center of the field of view; hence a diagonal symmetry plane (here NW–SE) is present. Also, if the last bit of the isogyre is seen along a diagonal direction that passes through the center of the field of view (the dashed line in Fig. 8-40A), it follows that when the 45° position is reached, the melatopes will still be located along that diagonal, a

Figure 8-41.
Monosymmetric Bxo interference figures. (A) As the isogyres leave the field simultaneously during stage rotation, the last bit of darkness is *not* seen on a diagonal line (dashed). This indicates that a single NW–SE symmetry plane is present at the 45° position. (B) The isogyres leave at different times during stage rotation, but on the diagonal. This indicates that a single NE–SW symmetry plane is present at the 45° position.

condition that indicates that a second (here NE–SW) symmetry plane is present (Fig. 8-40B). The figure is established as being bisymmetric.

Alternatively, if the isogyres leave at unequal times, or are last seen at locations other than on a central diagonal direction, the figures are either monosymmetric or nonsymmetric. Examples of such situations are shown in Figure 8-41. Note that isochromes may give further clues on figure orientation.

DETERMINATION OF OPTIC SIGN

The optic sign of a biaxial material is easily determined with the aid of interference figures and accessory plates. It is only necessary to determine the type of interference figure being viewed, to rotate the figure into a known orientation, and then to interpret the color change upon insertion of an accessory plate. In some cases, where the color reactions are vague in the interference figure, it may be necessary to use the interference figure to indicate the vibration directions of the grain, and then to determine the sign with orthoscopic illumination. The trick in all of this is to be sure that the type of interference figure is identified correctly.

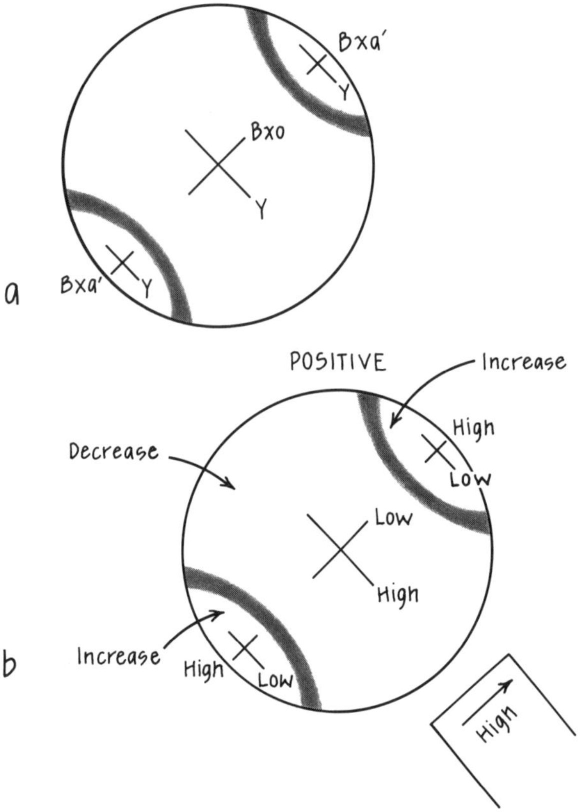

Acute Bisectrix Interference Figures

Consider a schematic bisymmetric Bxa figure, with the optic plane oriented in the NE–SW diagonal direction (Fig. 8-42A). The vibration directions between the isogyres, although variable in direction and index, can be characterized as Bxo oriented NE–SW and Y oriented NW–SE (see Fig. 8-27). Highly inclined rays emerging at the concave sides of the isogyres can be characterized as having Bxa' vibrations (NE–SW) and Y vibrations (NW–SE) (Fig. 8-30).

If the mineral is optically positive, then Bxa = Z (γ) and Bxo = X (α). Between the isogyres (Fig. 8-42B), the Y (β) vibration has a higher index than that of Bxo (α). Outside of the isogyres, on the concave side, the Bxa' (γ') vibration has a higher index than that of Y (β). The high and low vibration directions are reversed on opposite sides of the isogyres. An accessory plate has its high index (slow) vibration direction oriented NE–SW. As this coincides with the high index direction on the concave sides of the isogyres, inserting the accessory will add the grain and accessory retardations in these areas. Between the isogyres, unlike vibration directions are parallel, resulting in a subtractive relationship between the grain and accessory retardations.

Figure 8-43 and Plate 2 (Figs. 3, 4, and 5) show sign determinations using both the first-order red plate and the quartz wedge. In part A the Bxa figure shows no color bands, only a white to gray field. Insertion of the red plate with a positive mineral results in addition (red plus white yields blue) on the concave side of the isogyres, and subtraction (red minus white yields yellow) on the convex side. These are, of course, the same type

Figure 8-42.
(A) Vibration types in a Bxa interference figure. (B) In a *positive* biaxial material, Bxa is γ and Bxo is α. Consequently, on the concave side of the isogyres the high vibration direction is NE–SW, as $\gamma > \beta$. Insertion of an accessory in standard orientation increases retardation in these areas. On the convex side of the isogyres, the high vibration direction is oriented NW–SE, as $\beta > \alpha$. The accessory decreases retardation in this area.

Figure 8-43.
Sign determination with a Bxa interference figure. (A) If isochromes are absent and the field is mainly white, use the first-order red plate or the mica plate. With the first-order red plate, an increase in retardation yields blue or blue-green; a decrease in retardation results in yellow. With the mica plate, decrease in retardation is characterized by the formation of a black spot adjacent to the isogyre; an increase is generally not apparent. (B) When isochromes are present, use the quartz wedge. During insertion, an increase in retardation is seen as movement of isochromes toward the optic axes. Isochromes move away from the optic axes when retardation is decreased.

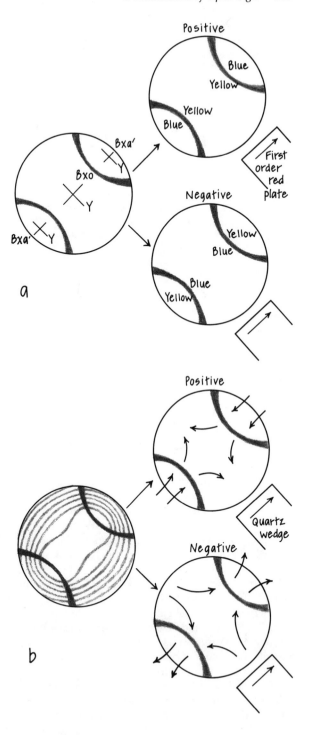

of color reactions that occur with a first-order red plate when a grain is viewed orthoscopically. A material having a negative sign reverses the color reaction.

When isochromatic color bands are present (part B), the sign is more easily determined by inserting the quartz wedge. An increase in retardation (during insertion) results in the isochromes moving toward the melatopes as higher level colors replace lower level colors. A decrease in retardation is seen by movement of isochromes away from the optic axes.

Obtuse Bisectrix Interference Figures

Sign determination on a Bxo figure is more difficult than on a Bxa figure, because the isogyres are not in the field of view when the figure is in a diagonal orientation. Figure 8-44A shows the general vibration directions and color arrangement when the optic plane is oriented NE–SW. As seen in part B, when the mineral is positive and Bxa (γ) is parallel to the high-index direction of the accessory, the colors rise and move in the directions indicated by the arrows. Movement is reversed for a negative mineral. Usually a quartz wedge is used for this determination, as the interference colors in the figure are commonly at least of second order.

110   Chapter 8: Biaxial Materials and Light

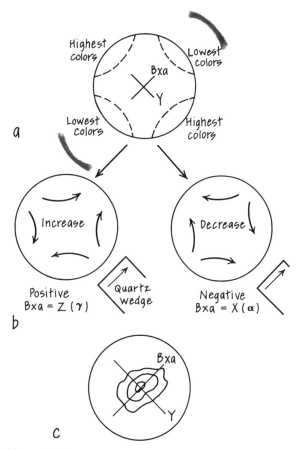

Figure 8-44.
Sign determination with a Bxo interference figure. (A) The figure has been rotated into a known orientation. The isogyres left the field in the NE and SW quadrants, indicating that Bxa is NE–SW and Y is NW–SE. The lowest interference colors are in the NE and SW quadrants, the highest in the NW and SE quadrants. (B) As colors are generally second order or higher, the quartz wedge should be used. In a positive mineral, Bxa > Y; during insertion, retardation increases in the entire field of view. Thus colors move inward in the NW and SE quadrants and outward in the NE and SW quadrants. A negative mineral shows opposite movement of colors. (C) If the color movement is obscure, convert to orthoscopic illumination (with crossed nicols). Vibration directions within the grain are known from the interference figure in part (A). Insertion of the quartz wedge will indicate whether Bxa is higher or lower than Y and thus establish the sign.

In many instances, the movement of colors is difficult to interpret. One answer to this is to rotate the stage until both sides of the isogyres are visible (just before they have left the field), then insert the accessory. Comparison of the opposite sides of the isogyres may reveal differences in color, thus leading to sign determination.

Another alternative is to use orthoscopic illumination. Rotate the stage until the figure is as shown in part A; this can be verified by noting that the isogyres have left the field of view in the NE and SW quadrants. Now view the grain orthoscopically by removing the Bertrand lens and the condenser (Fig. 8-44C). The vibration directions within the grain are identical to those in the center of the interference figure (part A), with Bxa oriented NE–SW and Y oriented NW–SE. Insertion of an accessory plate will reveal whether the Bxa has a higher or lower index than β, and the sign is thus determined.

Optic Normal Interference Figures

Direct determination of the optic sign with an optic normal figure is often useless because of the difficulty in interpreting the color reaction. It is better to use the interference figure to establish vibration directions within the grain and determine the sign orthoscopically.

Rotate the stage until the grain is in a diagonal orientation, such that the isogyres have left the field in the NE and SW quadrants (Fig. 8-45); this puts the Bxa vibration direction NE–SW and the Bxo vibration direction NW–SE. View the grain orthoscopically, then insert an accessory. If Bxa is higher in index than Bxo, the mineral is positive; if lower, the mineral is negative.

Optic Axis Interference Figures

Occasionally a biaxial mineral is found that remains in extinction. This occurs when one of the two optic axes is vertical. The figure consists of a single straight or curved isogyre that extends through the center of the field (Fig. 8-35). Iso-

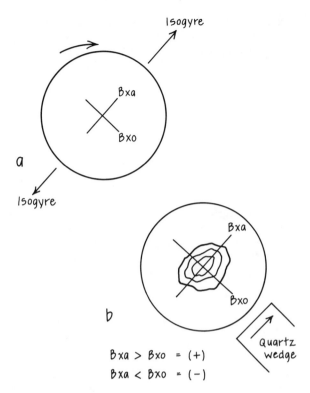

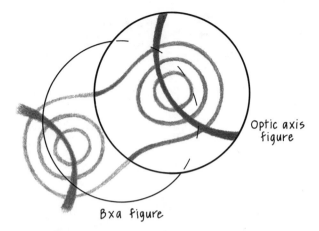

Figure 8-45.
Sign determination for an optic normal interference figure. (A) Rotate the stage until the isogyres leave the field in the NE and SW quadrants. This establishes the Bxa direction as being NE–SW; Bxo is NW–SE. (B) Convert to orthoscopic illumination (with crossed nicols) and insert the quartz wedge. If the retardation increases (outward color movement), the sign is positive. Decrease in retardation indicates a negative sign.

Figure 8-46.
An optic axis interference figure (heavy circle) can be regarded as a portion of a Bxa figure. Sign reactions are the same as those for the corresponding portion of the Bxa figure.

chromatic curves are concentric to the melatope, which is located at the cross-hair intersection. The optic axis figure can be regarded as being half of a Bxa figure (Fig. 8-46). In the example shown, it is obvious that the optic axis figure consists of the NE portion of a Bxa figure. The color changes, upon insertion of accessory plates, are the same as those shown for the NE portions of the diagrams in Figure 8-43.

The only problem that might arise is with materials that have a high $2V$. With them, it may be difficult to decide which side of the isogyre is convex and which is concave, that is, whether one is looking at the NE or the SW portion of a Bxa figure. If this cannot be determined, it means that the isogyre is effectively a straight line, and the $2V$ is essentially 90°. The sign is therefore indeterminate and the mineral is either positive or negative.

COMMON PROBLEMS

How is the sign determined with a mineral that consistently yields an interference figure that looks like a uniaxial inclined optic axis figure? During stage rotation, a single more-or-less straight NS- or EW-oriented isogyre moves across the field as shown in Figure 6-10 (p. 64). The mineral could be either uniaxial, or biaxial with a small $2V$; if biaxial, two orientations are possible (Fig. 8-47). Fortunately, the sign reaction is the same for the three possibilities. Rotate the stage until the isogyres have left the field of view to the north and east. If the mineral is either uniaxial positive or

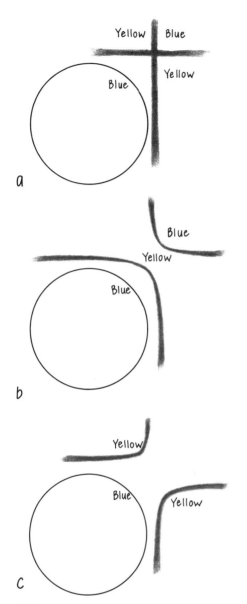

Figure 8-47.
If the melatope is located outside of the field of view, it is sometimes difficult to determine whether the grain is uniaxial (A) or biaxial with a small $2V$ (B and C). The sign, however, can be determined. Here, insertion of a first-order red plate (over a white field with no isochromes) has produced an increase in retardation (blue) in the NE portion of the field of view. This establishes that the substance is positive. A yellow color indicates a negative sign.

biaxial positive, the interference colors are additive when an accessory plate is inserted, and if the mineral is negative, the interference colors are subtractive. In the three possibilities shown, the symmetry of the figure indicates that a principal index direction is oriented NW–SE and parallel to the stage. Unfortunately, it is not known whether the index is ω (case A), Y (case B), or Bxo (case C). A decision must be made with other information from different grains.

How can one determine the optic sign of a biaxial mineral when it is not clear whether the interference figure is a Bxo or O.N. type? This is no problem. In both cases, rotation of the stage will cause the isogyres to leave the field in the quadrants that contain the Bxa vibration direction. Consequently the sign determination reaction is the same for both types of figures.

How does one determine the sign when it is not clear whether the interference figure is of the Bxa or Bxo type? The sign cannot be determined if these two types of figures cannot be differentiated. When this situation occurs, it usually means that the $2V$ is quite large. It is, therefore, necessary to determine the sign on a grain that yields an optic axis figure.

What if the mineral grain size is too small to yield interference figures? Several possibilities exist: (1) When working with thin sections, try to use the associated or host minerals as a clue to the nature of the unknown mineral; if the sample is pulverized, obtain a coarser sample. (2) Use information gained from cleavages, extinction behavior, and index measurements to limit the possibilities; this type of approach is discussed in Chapter 10. (3) Identify the mineral with an X-ray powder diffraction pattern or one of the other high-technology methods.

If the interference figure is a single isogyre that moves across the field during stage rotation, how does one determine whether the crystal is uniaxial or biaxial? Rotate the stage until the isogyre passes through the cross-hair intersection (Fig. 8-48). If, in this orientation, the isogyre does not coincide with the NS or EW cross hair, the crystal definitely is biaxial. If the isogyre is parallel to either the NS or EW cross hair, the crystal may be uniaxial *or* biaxial; interference figures from crystals in other orientations will be necessary to make the distinction.

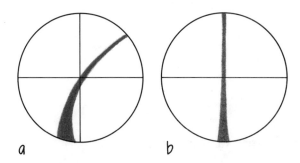

Figure 8-48.
Determination of optic character with highly off-center interference figures. Rotate the stage until an isogyre passes through the cross-hair intersection. (A) If the isogyre is not parallel to either cross hair, the substance is biaxial. (B) If the isogyre is always parallel to one of the cross hairs, the substance may be either uniaxial or biaxial.

ADDITIONAL READINGS

Kamb, W. B. 1958. Isogyres in interference figures. *American Mineralogist 43*, 1029–1067.

Slawson, C. B., and A. B. Peck. 1936. The determination of the refractive indices of minerals by the immersion method. *American Mineralogist 21*, 523–538.

Tobi, A. C. 1956. A chart for measurement of optic axial angles. *American Mineralogist 41*, 516–519.

Willard, R. J. 1961. A re-examination of some mathematical relations in anisotropic substances. *Transactions American Microscopical Society 80*, 191–203.

Wright, F. E. 1951. Computation of the optic angle from the three principal refractive indices. *American Mineralogist 36*, 543–556.

9

Optical Behavior of Biaxial Materials as Related to Crystal Symmetry

In our sad condition our only consolation is the expectancy of another life. Here below all is incomprehensible.
MARTIN LUTHER

It is common to think of the symmetry of a crystal only in terms of the orientation and number of related faces and cleavages. For example, using the Hermann-Mauguin symmetry symbol $4/m\ 2/m\ 2/m$ of the ditetragonal-dipyramidal class of the tetragonal system, a standard procedure is to relate the occurrence of a (100) crystal face to the presence of three equivalent faces ($\bar{1}$00), (010), and (0$\bar{1}$0). And it is but a small step to decide that if a good cleavage exists parallel to (100) and ($\bar{1}$00), a second equivalent cleavage exists parallel to (010) and (0$\bar{1}$0).

However, *all* of the directional properties of a crystal must conform to its crystallographic symmetry, not just the number and positions of crystal faces and cleavages. These include such properties as electrical conductivity, thermal expansion and contraction, solubility, hardness, and resistivity. Crystallographic restrictions on the positions of symmetry planes and axes of rotation also apply to the arrangement of optical parameters.

In this chapter, the constraints placed upon the orientation of the biaxial indicatrix in the orthorhombic, monoclinic, and triclinic systems will be examined; it will be seen that each crystal system has its own constraints relative to optical orientation. This, in turn, means that relations exist between the orientation of the indicatrix and that of possible cleavage surfaces. Knowledge of such relations can greatly ease the microscopic identification of unknown crystalline materials.

OPTICAL ORIENTATION OF ORTHORHOMBIC MATERIALS

The maximum symmetry of an orthorhombic crystal is that of the dipyramidal class, which can be characterized by the Hermann-Mauguin symbol $2/m\ 2/m\ 2/m$. Translated, this means that each of the three crystallographic axes is a two-fold axis of rotation, perpendicular to which is a plane of symmetry (Fig. 9-1A). This arrangement results in three mutually perpendicular symmetry planes, whose lines of intersection are twofold axes of rotation.

Figure 9-1B shows the symmetry of a biaxial indicatrix. Again, three mutually perpendicular symmetry planes are present, whose lines of intersection are two-fold rotation axes. As the crystallographic and optical symmetry must correspond, the three crystallographic reference axes must be parallel to the three principal optical directions. This permits six different possible correspondences, each of which is listed in Table 9-1. Figure 9-2 shows two examples.

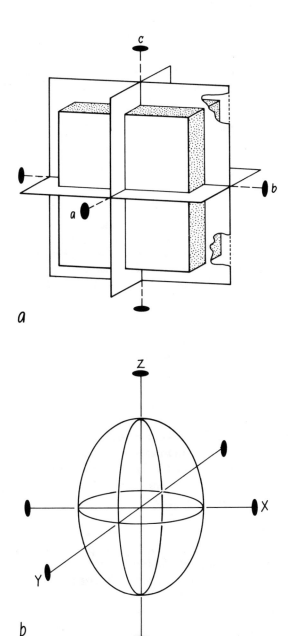

Figure 9-1.
(A) Symmetry of the dipyramidal class of the orthorhombic system consists of three perpendicular mirror planes, the intersections of which correspond to twofold axes of rotation. (B) Symmetry of a biaxial indicatrix consists of three perpendicular mirror planes, the intersections of which correspond to twofold axes of rotation.

Table 9-1.
Optical orientation of orthorhombic crystals

Crystallographic axes			Miller Index of optic plane (XZ)
a	b	c	
X	Y	Z	(010)
X	Z	Y	(001)
Y	X	Z	(100)
Y	Z	X	(100)
Z	X	Y	(001)
Z	Y	X	(010)

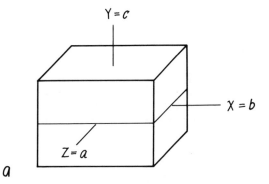

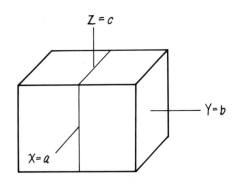

Figure 9-2.
Two examples of optical orientation in the orthorhombic system. The three principal optical directions must correspond in some manner to the three crystallographic axes. (A) $X = b$, $Y = c$, and $Z = a$; the optic plane (XZ) is parallel to (001). (B) $X = a$, $Y = b$, and $Z = c$; the optic plane is parallel to (010).

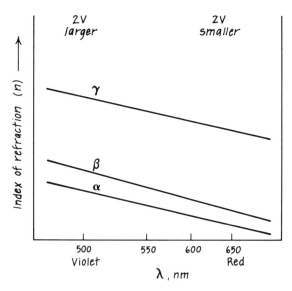

Figure 9-3.
Index dispersion results in different values of 2V for different wavelengths of light. In this example the dispersion curves of α and γ are parallel. The β index approaches α with increase in wavelength. Consequently, the 2V for red light is slightly smaller than the 2V for violet light.

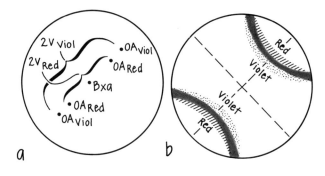

Figure 9-4.
Orthorhombic dispersion of the optic axes in polychromatic light. (A) Bxa is normal to the stage and Bxo is NE–SW. The 2V for red light is smaller than the 2V for violet light, thus the angular distance between the melatopes for red light (labeled OA_{red}) is smaller than the angular distance between the melatopes for violet light (labeled OA_{viol}). (B) In a Bxa figure, dispersion is seen as a violet hue on the convex sides of the isogyres and a red hue on the concave sides. Symmetry planes of the figure (dashed lines) are not changed by the presence of the dispersion fringes.

Dispersion

Within the rigidly oriented framework of orthorhombic symmetry it is possible to have index dispersion, similar to that described earlier (p. 16) as the change of index of refraction as a function of wavelength. Figure 9-3 shows the dispersion curves of a hypothetical orthorhombic mineral. The three indices decrease in value with increasing wavelength. As β is closer to α than it is to γ, the mineral is biaxial positive. Notice that α and γ decrease in index at the same rate, whereas β approaches α as the wavelength increases. The consequence of this is that the 2V (and 2E) of the mineral is smaller for red light than for violet light (Fig. 9-4A). *Dispersion of the optic axes* has been produced by index dispersion.

If an interference figure were produced by using red light, no red light would be present at the melatopes that mark the locations of the red optic axes; the isogyres would extend through these locations. If an interference figure were produced by using violet light, no violet light would be present at the melatopes for the violet optic axes. Consider the effect of using a normal polychromatic light source to produce an interference figure for this mineral. Where the isogyres for red light are located, no red light is present; consequently, violet light will be seen there. At the violet isogyre locations, no violet light is present, but instead, red is seen.

These colors show up as fringes on either side of the isogyres, with the strongest effects near the optic axes (Fig. 9-4B). This type of dispersion is described in terms of the 2V for red versus violet light. Here the 2V of red is less than that of violet, so in mineral descriptions, this would be listed as "Disp. r < v". Note well that when the red dispersion fringe is on the concave side of the isogyres, the 2V for red light is *less* than that for violet. The reverse condition, a smaller 2V for violet, would be described as "Disp. v < r".

The two diagonal symmetry planes of the Bxa figure (dashed in Fig. 9-4B) are not eliminated by the presence of this type of dispersion. This dispersion pattern establishes that the mineral under observation is orthorhombic. In monoclinic and triclinic minerals, dispersion patterns do decrease the symmetry of the interference figure. Do not assume, however, that dispersion of the optic axes is a universal panacea for mineral identification. Most minerals do not show distinct dispersion patterns, and even when present they may be difficult to interpret. Nevertheless, dispersion may sometimes be very useful.

Cleavage

The relationship between orthorhombic cleavages and the position of the three principal optical directions is generally much more important than dispersion of the optic axes. The various orthorhombic cleavage types are listed in Table 9-2 and shown in Figure 9-5.

A *pinacoidal cleavage* produces a set of surfaces that are parallel to a single planar direction within the crystal. For example, a {100} cleavage produces surfaces parallel to (100) and to ($\bar{1}$00). The cleavage fragments tend to lie on these surfaces

Table 9-2.
Relationship between orthorhombic cleavages and symmetry of interference figures obtained from cleavage fragments

Cleavage type	Miller index	Number of different cleavage directions	Symmetry of interference figures
Basal pinacoid	{001}	1	Bisymmetric
Front pinacoid	{100}	1	Bisymmetric
Side pinacoid	{010}	1	Bisymmetric
First-order prism	{0kl}	2	Monosymmetric
Second-order prism	{h0l}	2	Monosymmetric
Third-order prism	{hk0}	2	Monosymmetric
Dipyramid	{hkl}	4	Nonsymmetric

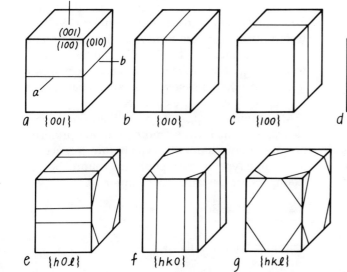

Figure 9-5.
Block diagrams illustrating the seven different types of orthorhombic cleavages (Ehlers, 1980). Pinacoidal cleavages (A, B, and C) produce a single direction of cleavage. Prismatic cleavages (D, E, and F) produce two directions of cleavage, and dipyramidal cleavage (G) produces four directions of cleavage (Ehlers 1980, Fig. 1).

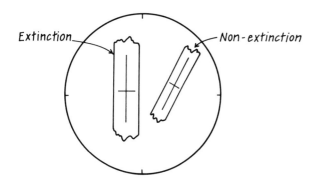

Figure 9-6.
Orthorhombic materials with a prismatic cleavage yield elongate grains with parallel extinction. A principal index direction is parallel to the direction of elongation.

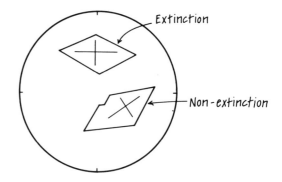

Figure 9-7.
Pyramidal cleavage in the orthorhombic system generally yields parallelogram-shaped grains that have symmetrical extinction.

but present an irregular outline when viewed through the microscope, as only a single direction of cleavage is present. As the orthorhombic pinacoids are perpendicular to the crystallographic reference axes, a cleavage fragment lying on a pinacoidal surface is oriented such that one of these axes is perpendicular to the stage. As each crystallographic reference axis corresponds in direction to one of the three principal optical directions, either X, Y, or Z is perpendicular to the stage. Consequently, either a bisymmetric Bxa, Bxo, or O.N. interference figure is obtained. When examining orthorhombic grain fragments, the repeated occurrence of bisymmetric interference figures is indicative of a pinacoidal cleavage.

Prismatic cleavages ({0kl}, {h0l}, and {hk0}) produce two directions of cleavage parallel to a single crystallographic reference axis (a, b, or c, respectively). Cleavage fragments are prismatic in shape. Fragments lie on one of the cleavage surfaces, and the second cleavage appears in the microscope as an elongate edge (Fig. 9-6). As the fragments are elongated parallel to a crystallographic axis (and consequently a principal vibration direction), they exhibit parallel extinction.

The interference figure produced by a grain lying on a prismatic cleavage is monosymmetric. Perpendicular to the single plane of symmetry within the figure is a principal vibration direction which is parallel to the stage; this vibration direction (X, Y, or Z) can be measured with a Becke test after bringing the long direction of the fragment in line with the permissible direction of the lower polarizer.

A *pyramidal cleavage*, {hkl}, produces four directions of cleavage, none of which is parallel to a crystallographic reference axis or principal vibration direction. Fragments as seen in the microscope have the general outline of a parallelogram[1] and show a symmetrical extinction (Fig. 9-7). Interference figures are nonsymmetric, as none of the three principal vibration directions is parallel to the stage. Fortunately, pyramidal cleavages in the orthorhombic system are rare; pinacoidal or prismatic cleavages are common.

Many minerals possess more than one type of cleavage. In such cases, more than one preferred orientation of fragments, type of interference figures, and type of extinction are found. Several examples are described in Chapter 10.

---

1. The number of edges seen could be six if the axial ratio approaches 1:1:1 (as in fluorite).

OPTICAL ORIENTATION OF MONOCLINIC MATERIALS

The maximum symmetry of the monoclinic system is that of the prismatic class—$2/m$. A single twofold rotation axis is parallel to the b crystallographic axis, and a perpendicular symmetry plane is parallel to the ac plane. This sets up the simple restriction that one of the three symmetry planes of the indicatrix is parallel to the ac plane. It follows that the principal vibration direction perpendicular to this optical symmetry plane is parallel to the crystallographic b axis (Fig. 9-8). The other two principal vibration directions are within the ac plane; neither, however, is necessarily parallel to the a or c crystallographic axes. Thus, we can think of the indicatrix as having one of its principal vibration directions coincident with the b axis, and the other principal vibration directions as having positions perpendicular to each other within the ac plane.

Dispersion

As with dispersion within the orthorhombic system, monoclinic dispersion is usually difficult to observe and generally of little practical use. Where it can be observed and interpreted properly, however, it may furnish valuable information.

As already noted, the only restriction on optical orientation is that the crystallographic b axis must coincide with a principal vibration direction. The amount of inclination of the other two principal vibration directions within the ac plane varies with mineral species. The orientation of a monoclinic mineral might be completely described by a statement such as "$b = X$ and $Z \wedge c = +34°$" (Fig. 9-9); the plus sign (usually omitted) is taken to mean that Z is inclined $34°$ in front of the positive end of the c axis. The third principal vibration direction, Y, would necessarily be oriented such that $Y \wedge c = -56°$.

The two principal vibration directions within the ac plane of a monoclinic material often have fixed angular relationships to the crystallographic a and c axes for all wavelengths of light. Some,

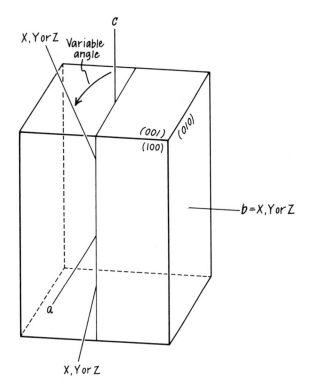

Figure 9-8.
Optical orientation in monoclinic crystals. The b crystallographic axis corresponds to one of the three principal vibration directions. The remaining two principal vibration directions are within the ac plane; orientation within this plane is generally given by stating the angle between the c axis and the larger of the two indices within the plane (for example, $Z \wedge c = 40°$).

however, have a slightly different amount of inclination about the b axis for different wavelengths of light, as permitted by the symmetry formula $2/m$. This results in *dispersion of optical orientation*, which may occur along with index dispersion. Dispersion of optical orientation results in color fringes about the isogyres, with the color arrangement being indicative of monoclinic symmetry. Three types of such dispersion are possible.

Crossed dispersion of the optic axes may occur when $b = Bxa$. It can be observed when combined with index dispersion; Figure 9-10A illustrates

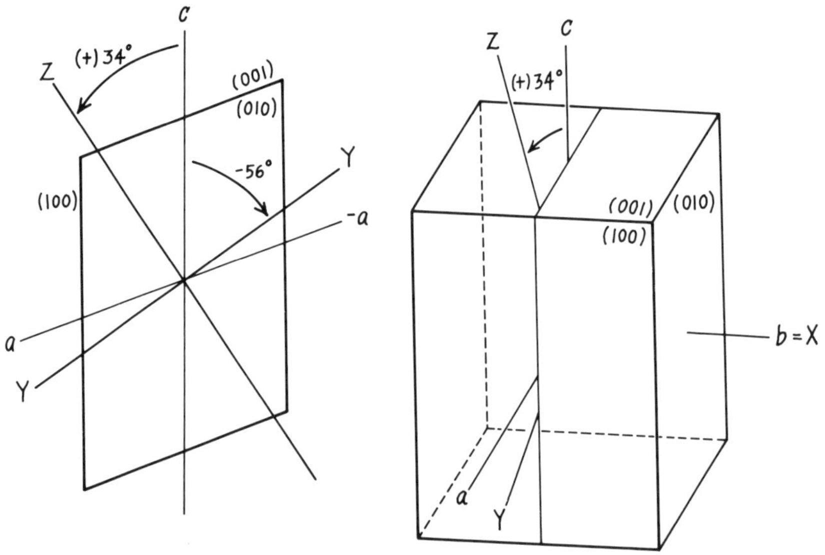

Figure 9-9.
A typical optical orientation of a monoclinic crystal: $b = X$ and $Z \wedge c = 34°$. Positive angles mean that a principal vibration direction is inclined between the positive ends of the a and c axes; a negative angle means that a principal vibration direction is inclined between $+c$ and $-a$. In this example $Y \wedge c = -56°$.

this case. The Bxo for red light (r) is inclined at a smaller angle to the c axis than the Bxo for violet light (v). The optic planes for red and violet light are shown as dashed lines. The vibration directions of intermediate wavelengths are inclined in intermediate amounts. Part B shows the positions of the two optic planes as viewed along Bxa. Included are the positions of the red and violet optic axes, with $2V_{red}$ greater than $2V_{violet}$ (that is, r > v). Part C shows the positions and types of color fringes that are developed adjacent to the isogyres with this type of dispersion.

The positioning of the color fringes eliminates both of the diagonal symmetry planes that would have normally been present in a centered Bxa figure. As noted earlier, orthorhombic dispersion did not eliminate symmetry planes within the interference figure; crossed dispersion, however, establishes an unknown material as being monoclinic. It follows that the single monoclinic symmetry plane must be parallel to the stage (as it is not present in the figure); hence the crystal is lying on (010), and Bxa coincides with b. Note, too (Fig. 9-10D), that lack of symmetry in the interference figure is also obvious when the figure is oriented in the cross position.

Inclined dispersion of the optic axes may occur if $b = Y$. The Y vibration is fixed, and Bxa and Bxo may have different inclinations (as a function of wavelength) within the ac plane as shown in Figure 9-11A. Part B shows the positions of the optic axes in a Bxa interference figure. As there are different orientations of Bxa as a function of wavelength, the violet and red acute bisectrices must be located along a line (here the NE–SW diagonal), rather than at a single point. If we assume no index dispersion, the optic axes for red and violet could be located as shown. The violet color fringe is on the concave side of one isogyre and the convex side of the other; the same is true for the red fringe (part C). If index dispersion were also present, it would be possible for the optic axes to coincide at one isogyre and have color fringes on the other.

The important thing to note about the interference figure in part C is that a single NE–SW symmetry plane is present. This plane corresponds to the single symmetry plane that is present in the ac plane of a monoclinic crystal. Thus, we have established with this interference figure that (010) is oriented NE–SW, that the crystal is monoclinic, and that $b = Y$ (as the Y vibration direction is perpendicular to (010))—a wealth of information from one interference figure.

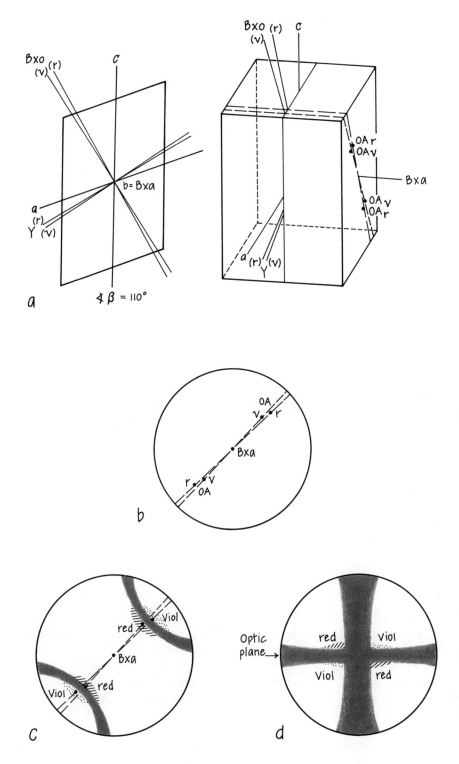

Figure 9-10.
Crossed dispersion of optic axes in the monoclinic system. (A) The *b* axis is parallel to Bxa. The Bxo for violet light has a greater inclination from the *c* axis (34°) than does the Bxo for red light (31°). Intermediate colors (and wavelengths) are inclined intermediate amounts. (B) When looking along the Bxa direction, the optic planes for violet and red light are seen to be at slightly different orientations about Bxa. Because of index dispersion, the 2V for red light here is greater than that for violet light; the optic axes (O.A.) for red light are more widely separated than those for violet light. (C) A Bxa interference figure, when dispersion is r > v, has red fringes on the convex side of the isogyres (violet optic axes) and violet fringes on the concave sides (red optic axes). The color fringes are not symmetrical across NW–SE or NE–SW vertical symmetry planes; crossed dispersion eliminates both symmetry planes. (D) Rotation of the stage 45° clockwise from (C) retains the nonsymmetric dispersion pattern.

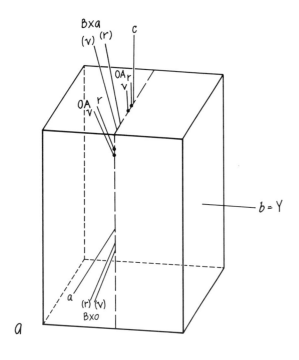

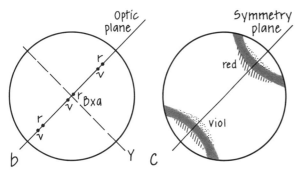

◀ **Figure 9-11.**
Inclined dispersion of optic axes in the monoclinic system. (A) The *b* axis is parallel to *Y*. The Bxa for violet light has a greater inclination from the *c* axis than does the Bxa for red light. The red and violet optic axes are similarly displaced; index dispersion is not present. (B) The displacement of the optic axes and acute bisectrices can be seen when viewed along an intermediate Bxa direction. (C) A Bxa figure shows red fringes on the SW sides of the isogyres and violet fringes on NE sides. The single symmetry plane that is present must be parallel to (010) (see Fig. 9-8); the optic plane is also established as corresponding to (010).

Figure 9-12. ▶
Parallel dispersion of optic axes in the monoclinic system. (A) The *b* axis is parallel to Bxo. The Bxa for red light has a greater inclination from the *c* axis (34°) than does the Bxa for violet light (31°). The red and violet optic planes intersect along the Bxo direction. In this example, the 2V for red light is smaller than that for violet light. (B) When viewed along an intermediate Bxa direction, more than one optic plane is seen. (C) A Bxa figure shows violet fringes on the convex side of the isogyres and red fringes on the concave sides (as v > r). A single NW–SE symmetry plane is retained that must correspond to (010). It follows from this that Bxo = *b*. (D) The grain in part (C) rotated clockwise to the isogyre cross position. The single symmetry plane (010) is now NS.

Parallel dispersion of the optic axes may be developed when *b* = Bxo. In order to observe color fringes, it is also necessary to have index dispersion. Figure 9-12A shows the effect of dispersion of the optic axes. Different angles of tilt for the red and violet vibration directions of Bxa and *Y* are shown within the *ac* plane. The optic planes for red and violet light are indicated by dashed lines. Part B shows the positions of the optic axes relative to the acute bisectrices, assuming the case of r < v. The positions of the color fringes are seen in the interference figure in part C.

The interference figure in part C shows that a single symmetry plane is retained. This plane must correspond to the symmetry plane (010) in the monoclinic system. Hence the interference figure indicates that the (010) plane within the crystal is NW–SE, that the crystal is monoclinic, and that b = Bxo.

Note that the decrease in symmetry of the interference figure due to dispersion color fringes has only a very minor effect on index measurements. Dispersion should be disregarded when analyzing a figure for symmetry if measurement of indices is the primary goal.

Cleavage

Cleavage types for monoclinic minerals are listed in Table 9-3 and shown in Figure 9-13 (page 123). Recall that the only restriction on optical orientation is that the b crystallographic axis must coincide with one of the three principal vibration directions. As the {010} cleavage surface is perpendicular to the b axis, it follows that a cleavage fragment lying on a {010} surface must have a principal vibration direction perpendicular to the stage. The other two principal vibration directions can be measured, and a bisymmetric interference figure is obtained.

Cleavage fragments lying on either {100}, {h0l}, or {001} have the b axis parallel to the microscope stage. As the b axis coincides with a principal vibration direction, such grains yield a monosymmetric interference figure.

The remaining cleavages—{0kl}, {hk0}, and {hkl}—are not parallel to any of the principal vibration directions and thus yield nonsymmetric interference figures.

OPTICAL ORIENTATION OF TRICLINIC MATERIALS

The maximum symmetry of the triclinic system is that shown by the pinacoidal class—$\bar{1}$. Only a center of symmetry is present. Consequently there are no directional restrictions on the orientation of the indicatrix. Any orientation is possible.

Index dispersion as well as dispersion of optical orientation is shown in Figure 9-14A. The prin-

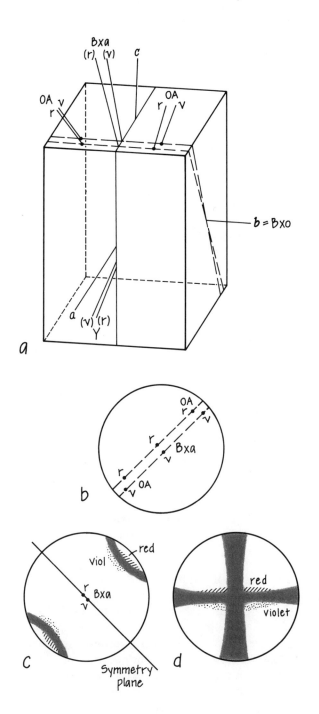

Table 9-3.
Relationship between monoclinic cleavages and symmetry of interference figures obtained from cleavage fragments

Cleavage type	Miller index	Number of different cleavage directions	Symmetry of interference figures
Basal pinacoid	{001}	1	Monosymmetric
Side pinacoid	{010}	1	Bisymmetric
Front pinacoid	{100}	1	Monosymmetric
First-order prism	{0kl}	2	Nonsymmetric
Second-order pinacoid	{h0l}	1	Monosymmetric
Third-order prism	{hk0}	2	Nonsymmetric
Fourth-order prism	{hkl}	2	Nonsymmetric

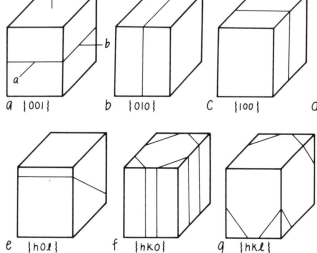

Figure 9-13.
Cleavages in the monoclinic system (Ehlers, 1980). Pinacoidal cleavages (A, B, C, and E) produce a single direction of cleavage. Prismatic cleavages (D and F) produce two directions of cleavage, as does hemi-dipyramidal (fourth-order prism) cleavage (G) (Ehlers, 1980, Fig. 8).

cipal optic directions and the optic axes for red light do not coincide with their equivalents for violet light. Part B shows a possible orientation of the optic axes and acute bisectrices that produce a Bxa interference figure in diagonal position. The resultant interference figure is seen to lack symmetry planes. Notice that the positions of the color fringes and the acute bisectrices in Figure 9-14C distinguish this figure from the nonsymmetric monoclinic figure in Figure 9-10C.

As a result of the freedom of orientation of the indicatrix, any triclinic cleavage fragment should yield nonsymmetric interference figures, as well as inclined extinction against any crystal faces or cleavage traces. In fact, though, some triclinic minerals are pseudo-monoclinic and can be treated as monoclinic. The extensive mineral descriptions in Volume 2 discuss the optical behavior of common triclinic materials when lying on cleavage surfaces. The triclinic feldspars are particularly good examples for study.

Additional Readings 125

ADDITIONAL READINGS

Ehlers, E. G. 1980. Use of cleavage as an aid in the optical determination of minerals, *Journal of Geological Education 28*, 176–185.

Louisnathan, S. J., F. D. Bloss, and E. J. Korda. 1978. Measurement of refractive indices and their dispersion. *American Mineralogist 63*, 394–400.

Stoiber, R. E., and S. A. Morse. 1972. *Microscopic Identification of Crystals*. New York: Ronald Press, 278 pp.

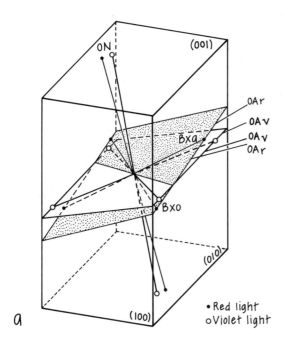

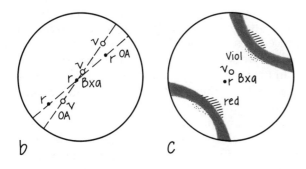

Figure 9-14.
(A) Minerals of the triclinic system may have both index dispersion and dispersion of optical orientation. (B) When viewed along an intermediate Bxa direction, optic planes for red and violet light have angular displacement in two directions. (C) The color fringes in a Bxa figure eliminate both symmetry planes. Note that this figure can be distinguished from the nonsymmetric monoclinic figure shown in Figure 9-10(C).

10

Microscopic Identification of Unknown Biaxial Materials

On with the dance, let joy be unconfined, is my motto; whether there's any dance to dance or any joy to unconfine.
MARK TWAIN

Identifying an unknown biaxial material is a simple task if all of the optical and crystallographic properties have been precisely determined. The determined properties are easily compared to descriptions of biaxial materials in determinative tables and charts. The second volume of this book describes the optical and crystallographic properties of more than 150 common minerals; tables and charts are arranged so as to be useful for determinations involving either thin-section or oil-immersion techniques. Note, however, that most minerals show considerable solid solution; as a consequence, the determined values may not exactly match a particular "book value."

Additional mineral descriptions by Heinrich (1965), Shelley (1985), and Kerr (1977) are mainly intended for use with thin-section analysis. Extensive descriptions of minerals for use primarily with oil-immersion techniques are given by Winchell and Winchell (1951, 1964), Troeger (1979), Phillips and Griffin (1981), Larsen and Berman (1934), and Fleischer, Wilcox, and Matzko (1984). General mineralogy is added to optical characteristics by Palache, Berman, and Frondel (1944, 1951, 1962) and Deer, Howie, and Zussman (1962, 1963, 1966, 1978, 1982). Books that stress sedimentary materials are Carozzi (1960), Cayeux (1970), and Milner (1952). These are all listed in the Bibliography. Each of these books takes a somewhat different approach to determinative procedures, and all should be examined if continued work in microscopy is anticipated.

In actual practice, it may be difficult or impossible to determine all of the optical characteristics of an unknown material. The reason may be time or equipment limitations, or an inherent factor in the material being examined. Certain materials may be too fine grained to yield interference figures or specific index determinations; other materials, because of perfect cleavage, may be limited to specific orientations; materials observed in thin section generally do not permit detailed estimation of indices of refraction. Thus it is often necessary to identify a material with incomplete data. Usually this is not a serious problem, as most of the common minerals can easily be identified on the basis of a few distinctive characteristics.

The following sections are intended to stress the fact that individual bits of optical and crystallographic information are often interrelated. When combined with each other, determinative procedures are considerably simplified.

IDENTIFICATION WITH OIL-IMMERSION TECHNIQUES

Interpretation of Published Data

Suppose that a variety of observations (say, positive optic sign, approximate 2V, and indices of refraction about 1.59–1.65) led you to suspect that an unknown material was the orthorhombic mineral *celestite*. The determination could be verified by a precise determination of α, β, and γ, but a quicker and easier method is to consider the consequences of the relationship between optic orientation and cleavage. The published optic orientation is $a = Z$, $b = Y$, and $c = X$; the cleavages are {001} perfect, {210} good, and {010} poor.

The approach is to superimpose this information on a pinacoidal block diagram. First sketch a block diagram showing the three pinacoidal forms and the crystallographic axis (Fig. 10-1A). Next, using the information on optic orientation, indicate the relationship between crystallographic axes a, b, and c, and the principal optical directions X, Y, and Z. Knowing that the sign is positive, it follows that $Z = $ Bxa, $X = $ Bxo, and Y, as always, is the optic normal (O.N.). The optic axial plane is parallel to {010}.

The various cleavage surfaces are added to the block. The {001} and {010} cleavages are pinacoidal, and therefore each is limited to a single orientation (see Table 9-2). The {210} cleavage is prismatic, and therefore has two orientations, one parallel to (210) and the other parallel to ($2\bar{1}0$). From the relationship between cleavages and principal optical directions it is possible to deduce the shape and orientation of cleavage fragments, the type of extinction, the type of interference figures, and finally, the indices of refraction that can be easily measured.

As the {001} cleavage is perfect, most of the fragments lie on that surface (gray area in Fig. 10-1A). The outline (as viewed in the microscope) is controlled by the remaining cleavages. The good {210} cleavage tends to form a four-sided grain outline, whereas the poor {010} cleavage has little effect. The crystallographic data in Palache,

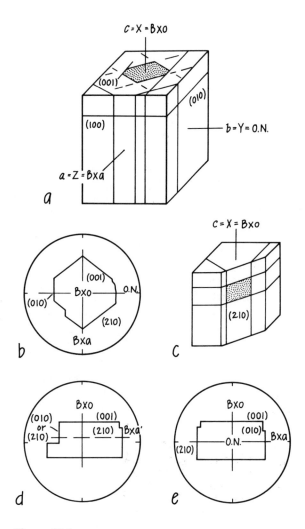

Figure 10-1.
(A) Pinacoidal block diagram of the orthorhombic mineral celestite. Optical and crystallographic directions are given, as well as cleavage traces. Darkened area indicates the outline of a {001} cleavage fragment. (B) A {001} cleavage fragment as seen in the microscope. (C) The darkened area indicates the outline of a {210} cleavage fragment. (D) A {210} cleavage fragment as seen in the microscope. (E) An uncommon cleavage fragment lying on the poor {010} cleavage as seen in the microscope (after Ehlers, 1980).

Berman, and Frondel (1951) reveal that the angles between {210} cleavage surfaces in celestite are 76° and 104° (Fig. 10-1B).

The {001} cleavage is normal to c and Bxo; therefore, the interference figure obtained from a grain lying on the {001} cleavage is bisymmetric Bxo. The two vibration directions parallel to {001} and the microscope stage are Bxa and O.N. As the mineral is positive, this permits estimation of γ and β. The vibration directions Z and Y bisect the angles between the (210) and ($2\bar{1}0$) cleavage edges; consequently these fragments have symmetrical extinction. If traces of the poor {010} cleavages are present, these are parallel to the Z vibration direction.

A lesser number of grains lie on the good {210} cleavage surfaces (gray area in Fig. 10-1C). The grain outline is elongate, as a result of the perfect {001} cleavage. A generally shorter grain edge at right angles to {001} is created by the {210} and {010} cleavages (Fig. 10-1D). The {210} cleavage surface is parallel to the c axis (Bxo) and consequently yields monosymmetric Bxa interference figures. The indices that can be estimated are α and γ'. As the vibration directions are parallel to the cleavage edges, the fragment has parallel extinction and will generally be length high (slow)—that is, the direction of elongation is parallel to (or at an acute angle to) the higher index direction (see also p. 135).

Few grains lie on the poor {010} cleavage surface. These yield a bisymmetric O.N. interference figure, permit estimation of α and γ, and have parallel extinction.[1]

Comparing the initial data and the information shown in Figure 10-1 with the data in determinative tables, it can be seen that the only mineral with which celestite can be confused (logically) is barite. The two are easily distinguished by making a mount with immersion liquid of index 1.63. The value of β, as measured on {001}, is lower than 1.63 for celestite and higher than 1.63 for barite.

The second volume of this book describes expected grain shapes, types of extinction and interference figures, and measureable indices of refraction on cleavage fragments for each of the described minerals. When additional data on dispersion, pleochroism, twinning, birefringence, optic angle, and color are considered, the identification procedure is considerably simplified.

Consider the type of information that can be extracted from a description of the monoclinic mineral *clinozoisite*. The optic orientation is given as $b = Y$, $c \wedge X = 0°-90°$; parallelism of only one crystallographic axis and principal optical direction verifies that the mineral is monoclinic. The angle between the crystallographic axes a and c, given as $\measuredangle \beta$, is 115°25'. The cleavages are {001} perfect and {100} good.

Again, begin with a pinacoidal block diagram (Fig. 10-2A). Superimpose the crystallographic axes and the principal vibration directions. As the inclination of c to X varies from 0° to 90°, X can be tentatively located at 45° in front of c (and parallel to the {010} plane). As $b = Y$, the third vibration direction Z must lie within the {010} plane at 90° to X. Traces of the pinacoidal cleavages {001} and {100} are added to the block; each has a single orientation. The mineral is biaxial positive, so Z = Bxa, X = Bxo, and Y = O.N.

As {001} is a perfect cleavage, most of the fragments lie on that surface. Elongation of the fragments results from the good {100} cleavage (Fig. 10-2A and B). The extinction is parallel, as the optic normal (Y) is parallel to both pinacoidal cleavages. As the optic orientation is highly variable, the vibration direction normal to grain length may be almost any value between Bxa (γ) and Bxo (α). Thus the fragments may be either length high

---

1. If the approach were reversed, such that the three types of cleavage fragments were of an unknown material, the fragments would show the material to be of orthorhombic symmetry. This is known for three reasons: (1) Normally only one type of cleavage in the monoclinic system, {010}, yields a bisymmetric interference figure; here two types of cleavage fragments yield them. (2) In the monoclinic system, a fragment with symmetric extinction normally has a monosymmetric rather than a bisymmetric figure. (3) The combination of fragments with both symmetric and parallel extinction is inconsistent with monoclinic symmetry.

Figure 10-2.
(A) Pinacoidal block diagram of the monoclinic mineral clinozoisite. The approximate optical directions and traces of the {001} and {100} cleavages are indicated. Darkened areas show the outlines of {001} and {100} cleavage fragments. (B) A {001} cleavage fragment as seen in the microscope. (C) A {100} cleavage fragment as seen in the microscope.

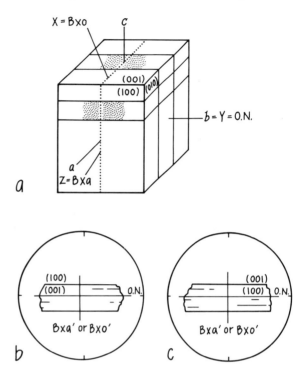

(slow) or length low (fast). The interference figure is monosymmetric Bxo or Bxa (as a function of the value of $X \wedge c$).[2]

A smaller number of grains lie on the good {100} cleavage surface (Fig. 10-2A and C). The fragments are elongate parallel to b as a result of the {001} cleavage. Extinction is parallel, as the O.N. (Y) is parallel to both pinacoidal cleavages. The grains are either length high (slow) or length low (fast) as a function of the variable optic orientation. The interference figure is monosymmetric Bxa or Bxo (as only the O.N. is parallel to the stage and the optic orientation is variable).

In summary, both types of fragments have parallel extinction, both are elongate parallel to Y, and both yield monosymmetric interference figures. Determination of the optic sign and an approximate β index of refraction would be sufficient to identify this mineral if it were an unknown substance.[3]

Consider the rare monoclinic mineral *colemanite*. The optical orientation is $b = X$ and $c \wedge Z = 84°$. The mineral is biaxial positive, has perfect {010} and poor {001} cleavage, and $\sphericalangle \beta = 110°$. This information is shown in the block diagram of Figure 10-3A. In Figure 10-3B a fragment is shown lying on the perfect {010} cleavage. The poor {001} cleavage, when present, creates an elongated grain outline. The fragment is elongated parallel to a as the {001} cleavage edge is parallel to a. The c axis makes an angle of 110° with a, as $\sphericalangle \beta = 110°$. As the {010} cleavage surface is normal to both b and Bxo, the interference figure is bisymmetric Bxo.

The extinction angle of this fragment is easily determined from the given information (Fig. 10-3C). Note the angles in the triangle formed by crystallographic axis c, the {001} cleavage edge, and principal vibration direction Z. The acute angle between c and the {001} cleavage is 70° (the supplement of $\sphericalangle \beta$). The angle between c and Z is given as 84°. As the three angles of a triangle must be 180°, the angle between Z and {001} must

---

2. In the unlikely situation that the angle between c and X is 25°25′, a fragment on {001} yields a bisymmetric Bxo interference figure and permits estimation of both β and γ. This situation occurs because angle β is 115°25′, which causes Bxa to be parallel to a.

3. The two types of cleavage fragments are a strong indication of monoclinic symmetry. Triclinic materials are unlikely, as grains lying on cleavage surfaces should generally have inclined extinction and nonsymmetric interference figures. If orthorhombic, it would be necessary to postulate two different prismatic cleavages, both parallel to the same optical direction, such as {110} and {120}. Such a combination is permitted by orthorhombic symmetry but is extremely uncommon.

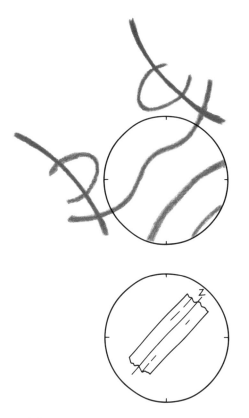

Figure 10-6.
Monosymmetric Bxa figure as seen in the 45° position. The plane of symmetry is NW–SE, and the principal index that can be estimated (Bxo) has a NE–SW vibration direction. The fragment that produced the interference figure is shown below in orthoscopic view. The NE–SW principal vibration direction of the figure coincides with the direction of elongation of the grain.

shown in part D, where $b = Y$; the two cleavages are oriented at a relatively small angle to X such that the fragments are all length high (slow).

The problem of identifying cleavage types and crystal system is resolved with the use of interference figures. In our example, figures from a variety of fragments turn out to be identical—monosymmetric Bxa (Fig. 10-6). During stage rotation the isogyres leave the field simultaneously, but not along the central field diagonal. Consideration of the symmetry of the figure reveals that Bxo (γ) is parallel to the stage and oriented NE–SW. Orthoscopic examination of the fragment (in the same orientation) reveals that Z is parallel to the length of the grain.

Consistency of interference figure orientation in a variety of grains indicates that both cleavages are crystallographically identical, that is, both of the same form. Of all the possibilities shown in Figures 10-4 and 10-5, only the orthorhombic prism (Fig. 10-4B) has two crystallographically equivalent cleavages parallel to Z.

Now that we know that the unknown material is of orthorhombic symmetry and has prismatic cleavage parallel to Z, it becomes a simple matter to complete the determination. Examination of determinative tables reveals that only two minerals fit our description, bronzite and hypersthene. Both are orthopyroxenes that fall within the enstatite-orthoferrosilite solid-solution series. A precise determination of the easily measured index γ reveals the composition and mineral name.[5]

For a final example, let us suppose that an unknown mineral has been determined to be biaxial positive, with indices in the general range of 1.68–1.75. Interference figures are identical in orientation and nonsymmetric, but the $2V$ can be determined to be about 40°–60°. The colorless fragments are elongated, and the higher index, γ', makes an extinction angle of about 36° with the grain edge.

It is obvious that at least two directions of cleavage are present; fragments lie on one cleavage, and the second forms the edge of the grain. The consistency in the interference figure's orientation indicates that the two cleavages are of the same crystallographic form.

The unknown material could not have orthorhombic symmetry, because cleavage fragments in this system produce only parallel or symmetric extinction. It is unlikely that the material is of triclinic symmetry, because within the triclinic sys-

---

5. In normal practice we would not set up all of the possibilities given in Figures 10-4 and 10-5 but would instead have examined the interference figures early in the determination.

tem, each of the two cleavages should yield interference figures of a different orientation. We are left with the monoclinic system.

Within the monoclinic system there are three different forms that yield two cleavage orientations. These are {hk0}, {0kl}, and {hkl} (see Table 9-3). Grains lying on any of these cleavage surfaces have inclined extinction, and the interference figures are nonsymmetric. Our unknown mineral matches all three cleavage categories.

At this point it is worthwhile to examine determinative tables for biaxial positive minerals that possess one of the three cleavage types. Between average β indices from 1.67 to 1.80 we find nine monoclinic minerals with {hk0} cleavage:

>Hornblende
>
>Diopside
>
>Omphacite
>
>Salite
>
>Riebeckite
>
>Pigeonite
>
>Augite
>
>Ferrosilite
>
>Aegirine-augite

All of these minerals are in the amphibole or pyroxene groups. The list can be culled by recalling that the material is colorless; this eliminates hornblende, salite, riebeckite, and ferrosilite. Our approximate $2V$ estimate of $40°-60°$ eliminates pigeonite ($2V = 0°-32°$) and aegirine-augite ($2V = 70°-90°$). We are left with diopside, omphacite, and augite, all of which are pyroxenes. Recourse to the detailed descriptions of these minerals will reveal the particular measurements or observations needed to complete the determination. The distinction among the three can be made on the basis of indices measured on cleavage surfaces, improved estimates of $2V$, or consideration of occurrence. The main point to remember is that a very small amount of information very rapidly narrowed the choices to only three minerals. With little extra effort, the particular mineral can be determined.

IDENTIFICATION IN THIN SECTION

The easiest and most hazardous way to identify a material in thin section is to do it on the sole basis of having seen it earlier in a different thin section. For the most part, a particular mineral tends to look the same in different sections. This sameness often applies to shape, color, relief, pleochroism, optic angle, associated phases, grain size, and the presence or absence of twins, cleavages, inclusions, alteration, etc. It is, however, very easy to make a mistake with this approach. Untwinned feldspars or cordierite are easily mistaken for quartz. Hornblende, when viewed in {100} section, shows little pleochroism and no cleavage, as does biotite when viewed in {001} section. Clinopyroxenes and olivines are often mistaken for each other because the microscopist has neglected to determine the optic angle or check for the presence of prismatic cleavage.

The way to get around this problem is simple. When learning to recognize a mineral, also spend a few minutes learning the look-alikes. There are usually very simple ways to distinguish between the presumed mineral and its look-alikes. Then, take the trouble to check the property that verifies the identity of the mineral that you "know"; occasionally you will be surprised to find that you have some uniaxial negative quartz (nepheline) or biaxial quartz (feldspar or cordierite).

If the mineral is not easily recognized by its obvious characteristics and mineralogical associations, it is necessary to be more systematic and determine a larger variety of properties.

Color and Pleochroism

Color is easily observed with uncrossed nicols. Colored minerals have their own determinative table in Volume 2. If the mineral is colored, determine if it is pleochroic. Using several grains (to

Figure 10-7.
Determination of a pleochroic formula. (A) Interference figures are obtained on several grains. The grain on the left yields a Bxa figure and that on the right an optic normal figure. The optic sign is determined as positive. (B) Orthoscopic view of the grains that produced the interference figures. (C) Clockwise and counterclockwise stage rotations bring the grains of part (B) into extinction, such that known vibration directions are now oriented EW. Uncross the nicols and determine the grain colors that are associated with each vibration direction. From this the pleochroic (absorption) formula can be written. In this case, Z yields the darkest color and greatest absorption, and Y yields the lightest color and least absorption. The absorption formula is $Z > X > Y$.

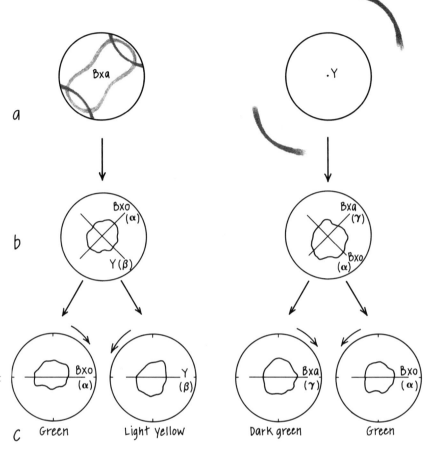

assure a variety of orientations), view the mineral with uncrossed nicols during stage rotation. If the grains reveal a change from one color to another, or a change in intensity of a single color, the mineral is pleochroic. Pleochroic minerals also have their own determinative table in Volume 2. As the differences in light absorption are directly related to the three principal optical directions (X, Y, and Z), it is desirable to establish the particular relationship between colors and optical directions for the mineral. First obtain identifiable interference figures and from them establish the vibration directions within the grain; orthoscopic examination of the corresponding grains will reveal the absorption color that is related to each vibration direction. An example of the approach is given in Figure 10-7.

Interference Colors

In a thin section, all grains are of the same thickness (ideally 0.03 mm). Consequently the observed variety of interference colors is dependent upon the birefringence and orientation of the particular minerals that are observed. For any single mineral the birefringence is fixed, and the observed interference colors are dependent upon orientation alone. Recall that the particular level of interference color that is observed depends upon the difference in velocity (and consequently difference in index of refraction) between the two vibration directions within the grain that are parallel to the stage. This relationship is summarized in Table 8-1.

Birefringence

It is often very useful to determine the birefringence of a mineral in thin section. As already noted, the level of interference colors (for grains of a single mineral) varies as a function of orientation. As the thickness is constant for all grains, the *maximum* level of interference color is dependent upon the mineral's birefringence.

In order to determine the birefringence of a mineral, it is necessary to examine a number of grains in different orientations (perhaps 10–15) and choose the one with the maximum interference color; this can be verified with an O.N. interference figure as Y is vertical. Using the maximum interference color and the grain thickness, birefringence can be determined with the aid of the Michel–Lévy chart (Fig. 7-2 and Plate 1).

Indices of Refraction

It will probably be necessary to determine the approximate indices of refraction of an unknown mineral. These can be estimated on the basis of relief, grains of lowest relief having indices matching that of the cementing material. It is a good idea to study the relief of some known minerals in order to have a comparison with unknown materials. Be sure to perform a Becke test between the unknown material and the mounting medium, to establish whether the unknown material has positive or negative relief (higher or lower index than the medium's, respectively). Check to see whether the indices of the unknown material are higher or lower than those of any known surrounding materials.

Cleavage and Habit

One of the unique advantages of thin sections is that grains can be observed in a variety of orientations. Consequently the number and angular relations among cleavage surfaces are easily observed. Observation of crystal habit is facilitated as well. It is but a simple step to relate cleavage directions to crystal habit and then, with the aid of interference figures and accessory plates, to relate these to principal optical directions. These relationships are extensively described in Volume 2.

Elongation Direction

Many minerals in thin section are elongate as a result of crystal habit, in contrast to crushed grains, whose elongation is a result of two or more cleavages. Grains in thin section can be classified according to the principal vibration direction that is either parallel or at an acute angle to the direction of elongation.

The principal vibration direction that is closest to the direction of elongation is easily determined with an accessory plate. First cross the nicols and rotate the grain into extinction (Fig. 10-8A), then rotate 45° from extinction (Fig. 10-8B) to place the vibration directions in diagonal orientation. Insert an appropriate accessory plate, and based on the change in interference colors on the grain, determine whether the vibration direction that is closest to the direction of elongation is a relatively high or low index. Do this for a number of grains. If the elongation direction of all of the grains is consistently high, this means that the vibration direction closest to the elongation direction is Z; such grains are called length high or length slow, or are said to have positive elongation. If the elongation direction of all of the grains is consistently low, this means that the closest vibration direction is X; such grains are called length low or length fast, or are said to have negative elongation. Grains that have the Y vibration direction closest to the direction of elongation are generally (as a function of orientation) both length high and length low (length slow and length fast), and possess both positive and negative elongation (Fig. 10-8C). This feature is described for each mineral in Volume 2.

Extinction Angles

As the a, b, and c crystallographic axes of an orthorhombic mineral are parallel (in any order) to X, Y, and Z, the extinction positions are generally parallel or symmetric to cleavage traces or crystal faces. In monoclinic minerals, only b is

136 Chapter 10: *Microscopic Identification of Unknown Biaxial Materials*

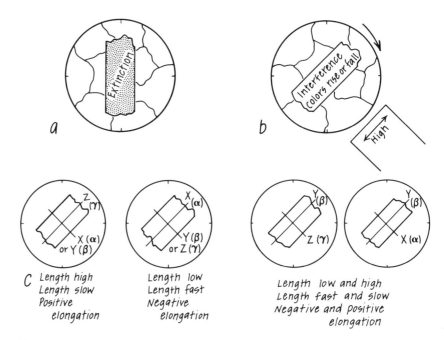

Figure 10-8.
Correlating elongation with principal vibration direction. (A) An elongate grain is rotated into extinction. (B) A clockwise rotation of 45° brings the vibration directions into diagonal orientation. The accessory plate raises or lowers the interference colors seen in the grain. An increase means that the grain is length high (slow), a decrease indicates the grain is length low (fast). (C) If a number of grains yield only a length high (slow) reaction, the vibration direction Z is parallel to (or less than 45° from) the elongation direction. Consistent length low (fast) reactions indicate that X is parallel to the elongation direction. If a variety of grains yield both reactions, Y is parallel to the elongation direction.

aligned with X, Y, or Z. Consequently, parallel or symmetric extinctions are expected only in those grains whose *b* crystallographic axis is parallel to the section. Other orientations yield inclined extinctions. Triclinic minerals, lacking a systematic relationship between crystallographic axes and optical directions, generally exhibit inclined extinction.

The extinction angle (the angle between a vibration direction and a crystal face or cleavage trace) varies as a function of grain orientation. It is often useful to determine the maximum extinction angle by checking a number of differently oriented grains. Try to determine the orientation of the grain with maximum extinction angle by noting the type of interference figure obtained, the grain shape, and the orientation of cleavage traces and twin planes. Such information can commonly be related to published block diagrams. A maximum extinction angle in the monoclinic system is often obtained on grains whose *b* axis is normal to the section.

Other Factors

A small number of minerals (listed separately in Volume 2) have anomalous interference colors; this provides an almost instant identification.

Twins of various types are common in biaxial minerals. If twinning is present, try to identify its type. It is often possible to relate the composition plane of the twins to cleavage surfaces, or crystallographic or principal optical directions.

Susceptibility to alteration, degree of euhedralism, the presence or absence of inclusions, and compositional zoning all contribute to identification of unknown materials.

Finally, one of the most important aspects of identification is the geological and mineralogical environment of the specimen. Many minerals are characteristic of a particular geological environment, such as glaucophane in high-pressure metamorphic environments. Certain minerals tend to be found together, such as nepheline and cancrinite. Other combinations are incompatible, such as nepheline and quartz. Recognition of alteration products helps to identify the host mineral. Knowledge of such petrogenetic relationships vastly simplifies the identification process, and provides considerable information on the origin of the rock.

In fact, after you have developed some expertise with the microscope, you might even enjoy using it. Good luck!

ADDITIONAL READINGS

Chamot, E. M., and C. W. Mason. 1958. *Handbook of Chemical Microscopy*, Vol. 1. New York: John Wiley & Sons, 502 pp.

Hutchison, C. S. 1974. *Laboratory Handbook of Petrographic Techniques*. New York: Wiley-Interscience, 527 pp.

Kennedy, G. C. 1947. Charts for correlation of optical properties with chemical composition of some common rock-forming minerals. *American Mineralogist* 32, 561–573.

McCrone, W. C., and J. G. Delly. 1973. *The Particle Atlas*, 2d ed., 4 vols. Ann Arbor: Ann Arbor Science (unpaged).

11

Special Techniques

If you build a better mousetrap, you will catch better mice.
GEORGE GOBEL

This chapter provides a brief introduction to two specialized microscopic techniques, the detent spindle stage and the universal stage. Many other supplementary devices and techniques are available as well.

One of these is the *heating stage,* a device that enables the microscopist to observe polymorphic inversions and melting relationships at high temperatures. With the use of the *diamond pressure cell,* minerals can be examined microscopically at ultrahigh pressures. The *interference microscope* splits the incident light, transmitting one portion through a grain and a second through the surrounding medium; the two rays are combined by a special objective, and the resultant retardation (seen as interference colors) provides a measure of the difference in indices of refraction between the two materials. *Ultraviolet* and *infrared microscopy* are used for specialized purposes.

THE DETENT SPINDLE STAGE

The detent spindle stage is a deceptively simple device that permits rotation of a crystal fragment about a horizontal axis (Fig. 11-1). Observations made using a spindle stage can add extreme precision to the determination of indices of refraction and optic angles. The conventional approach (when used with both temperature and wavelength control) yields indices to an accuracy of 0.002 and optic angles to 1.0°; with a detent spindle stage, indices can be determined to 0.0003 and optic angles to 0.1°. In addition, grain rotation permits indices of refraction to be determined and interference figures to be observed in a variety of orientations on the same fragment. Use of the spindle stage was first emphasized by Rosenfeld (1950), developed by Wilcox (1959), and vastly enlarged by Bloss (1978, 1981).

Setting Up

The first step is to mount a single mineral grain. Grains whose diameters are between 0.074 and 0.149 mm (100 to 200 mesh or 2.75 to 3.75 phi units) are suitable for this purpose. Take a common sewing needle and cut off its thicker end, then coat the tip with a thin layer of either colored nail polish or a mixture of four parts water-soluble carpenter's glue and one part crude blackstrap molasses. The latter adhesive is preferred as it can be softened (with the breath or by holding near a light bulb) if necessary to adjust the grain orientation.[1] Before the adhesive solidifies, the coated end of the needle is touched to the grain. The mounted grain should be checked under a

---

1. If a different adhesive is used, be sure that the adhesive is not soluble in the immersion liquid. Care should be taken that the adhesive does not mantle the grain; using colored nail polish makes such inadvertent mantling more obvious.

binocular microscope to be sure that it is centered on the axis of the needle and does not exceed the needle diameter. Ideally the grain should be checked before mounting, in order to be sure that it is not an aggregate of smaller grains, a multiple twin, highly altered, strained, and so forth. Slide the mounting needle into the hollow rotation axis of the detent spindle stage (Fig. 11-1). The rotation axis can be rotated through 180°. At 10° intervals the rotation arm clicks into a detent (indentation) on a tiny vertically mounted protractor. Smaller intervals can also be read on the protractor.

Rotate the microscope stage to the zero degree position and place the spindle stage on it in an approximate EW orientation, such that the mounted grain is centered on the cross hairs, and the rotational arm is to the east. Then tape the spindle stage to the microscope stage, ready for grain immersion in a calibrated liquid.

Grain immersion is accomplished by the use of an ordinary microscope slide to which two small (¼ inch long) support bars (made from cut-up paper clips) have been glued in parallel orientation. The two metal strips form a small well that can hold a few drops of immersion liquid. After covering the metal strips with a small cover slip, immersion liquid is fed into the well, and the slide is slid into the arms of the spindle stage, such that the grain is immersed. Changing the immersion liquid is done by removing the slide and replacing it with either a second slide containing a different liquid or the original slide after changing the liquid. If precise results are wanted, the liquid replacement should be carried out twice for each change of immersion liquid to avoid contamination from previously used liquid adhering to the grain-needle mount.

The rotation axis of the spindle stage must be oriented precisely EW. To accomplish this, place the rotational arm into the 0° position and then rotate the microscope stage counterclockwise to extinction; note the stage reading. Then turn the rotational arm into the 180° position and rotate the microscope stage clockwise to extinction; note the stage reading. The midpoint between the

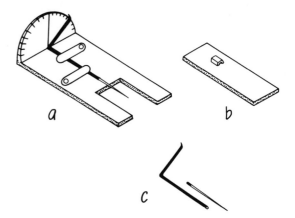

Figure 11-1.
(A) The detent spindle stage. One end of the axis of rotation (the spindle) fits against a protractor scale; the other end is hollow to allow insertion of a needle that holds the specimen. Detents along the circular edge of the protractor scale "click" the spindle into every 10° position. Ordinary glass slides (25 by 45 mm) fit snugly into the spindle stage's cut-out area. (B) Attached to a glass slide are two tiny rods bridged by a small cover glass. Immersion liquid is put in the space between the cover glass and slide. When inserted in the spindle stage, the mounted grain rests within the immersion liquid. (C) The spindle, removed from the stage, consists of appropriately bent stainless steel tubing. Standard sewing needles with mounted grains are inserted into the end of the spindle (Bloss, 1981, Fig. 1-5).

two stage readings locates the EW rest position of the spindle stage. This position (M_R) is the zero reference point for any later microscope stage rotations (called M_S).

Interference Figures

In order to observe interference figures it is necessary to use an objective with a numerical aperture of 0.65. Objectives with larger numerical apertures focus too close to the fragment under examination and may crush the cover glass.

It is instructive to examine both uniaxial and biaxial minerals with the spindle stage in order to observe the relationships between the orientation

140   Chapter 11: Special Techniques

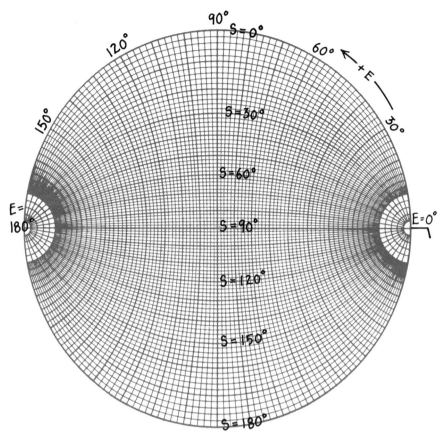

Figure 11-2.
The stereonet, oriented with its great circles east-west, represents the upper hemisphere. Each great circle represents a setting on the spindle axis (S). Thus, when the spindle axis is rotated 30° from the zero position, the great circle labeled $S = 30°$ is brought into horizontal orientation (the microscope stage). Values of E on the perimeter refer to the number of degrees of stage rotation that are necessary to bring the grain into extinction after appropriate rotation of the spindle (after Bloss, 1981, Fig. 1-9).

of grains and interference figures. These can be correlated with measurements of index of refraction and extinction angles.

Orthoscopic Observations

The optical character (uniaxial versus biaxial), the orientation, and the $2V$ of a crystalline fragment can be determined without recourse to interference phenomena with the aid of extinction curves.

The points that are plotted and joined to make up an extinction curve are obtained from the measurement of two rotational angles. One is the rotation on the spindle axis to a detent position, and the second is the rotation of the microscope stage to bring the crystal fragment into extinction. These angles are plotted on a standard stereonet (Wulff net), whose general characteristics are described in most introductory crystallography texts.

The stereonet is oriented so that its great circles run from east to west (Fig. 11-2). Each great circle corresponds to a setting on the spindle stage, the axis of which extends from the right side of the net. All points plotted (on an overlying sheet of tracing paper) will be located on the *upper hemisphere*, as is usual in crystallographic work. When the spindle is set at 0°, any point plotted on the tracing will be located on the outermost great circle, which is labeled as $S = 0°$. When the spindle axis is inclined so that the 30° detent is engaged, this brings the great circle labeled $S = 30°$ into the horizontal. A spindle setting of 90° brings the vertical great circle ($S = 90°$) into the horizontal.

Figure 11-3.
Constructing an extinction curve on the stereonet. Note first the rest position M_R of the microscope stage, which orients the spindle exactly EW; in this case $M_R = 5°$. With the spindle set at 0° ($S = 0°$) rotate the microscope stage clockwise to extinction. This occurs at a microscope stage reading M_S of 15°, thus the stage rotation to extinction E_S is $M_S - M_R$, or 10°. The second privileged vibration direction, always located at 90° from the first, must therefore have an extinction angle (E_S^{**}) at 100°. These two extinction points are plotted on the $S = 0°$ great circle at E values of 10° and 100°. The spindle is next set at 10° ($S = 10°$) and the stage rotated clockwise to extinction. An M_S of 22° means that the E_S value ($M_S - M_R$) is 17° and that E_S^{**} has a value of 107°. These two positions are plotted on the $S = 10°$ great circle. The process continues at successive 10° rotations of the spindle.

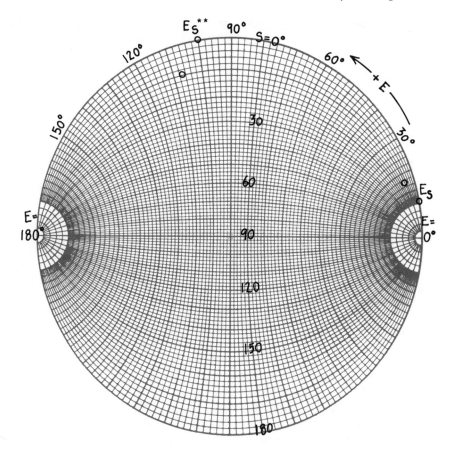

Consider the plotting of extinction positions on the stereographic projection. First set the spindle axis to 0°, then rotate the microscope stage from the rest position clockwise to extinction. When the crystal fragment becomes extinct, two privileged vibration directions are horizontal and oriented NS and EW. Assume in this case that the rest position (M_R) of the spindle axis is 5°, and the clockwise rotation has brought the grain into extinction at a microscope stage setting (M_S) of 15°. Extinction has been reached with a 10° clockwise rotation from the rest position. This means that the grain's perpendicular extinction positions at $S = 0°$ are located 10° counterclockwise from the rest position, as shown in Figure 11-3. The extinction angle for the microscope stage (E_S) is 10°. Note that two extinction points, 90° apart, have been plotted on the S_0 great circle as E_S and E_S^{**}.

The spindle is next set at 10° ($S = 10°$). Clockwise rotation of the microscope stage brings the grain into extinction at a setting of 22°. As M_S is 22° and M_R is 5°, the stage rotation required for extinction (E_S) is 17°. To plot this (Fig. 11-3), find the great circle corresponding to $S = 10°$; this is the plane that becomes horizontal at a spindle setting of 10°. On this great circle, plot the two extinction positions. One of these (E_S) is located at 17° counterclockwise from the spindle rotation axis, and the second (E_S^{**}) is located 90° away from the first on the $S = 10°$ surface.

This same procedure is continued for successive 10° increments of the spindle stage, to the limit of rotation at 180°. The data are completely tabulated (on a form such as shown in Fig. 11-4)

Figure 11-4.
A method of recording data obtained from extinction measurements. The values apply to Figure 11-3.

Mineral _____ $M_R = 5°$

S	M_S	E_S	E_S^{**}
0°	15°	10°	100°
10°	22°	17°	107°
20°			
30°			
40°			

before plotting them on the tracing. Note again that the extinction angle E_S is obtained from the difference between M_S and M_R. The second extinction position E_S^{**} is obtained by adding 90° to E_S if it is less than 90° or by subtracting 90° from E_S if it is more than 90°. This procedure maintains a convention of using only positive numbers for E_S and E_S^{**} and restricting E_S to values between 0° and 180°.

The tabulated data and the extinction curves for a uniaxial mineral are shown in Figure 11-5. One of the curves (usually broken into two portions) passes through the point where E_S or E_S^{**} is zero, at the plotted position of the spindle axis; this is called the *polar extinction curve*. A second curve crosses the spindle equatorial plane (at E_S or E_S^{**} = 90°); this is the *equatorial extinction curve*. In this example the equatorial extinction curve is a great circle (as can be verified by rotating the tracing until the plotted curve overlies a corresponding great circle of the stereonet). The fact that the equatorial extinction curve is a great circle means that the mineral under examination is uniaxial. The pole to this great circle (which is 90° away from it) falls on the polar curve. This point, labeled OA, shows the location of the optic axis. The coordinates of this point are $S = 165°$ and $E_S = 35°$. Assuming that the lower polarizer of the microscope is EW, the ε index of refraction of this fragment can be determined with a Becke test when the spindle axis is at 165° and the microscope stage is 35° clockwise from the rest position (M_R); this orientation places the optic axis in a horizontal EW position, parallel to the privileged direction of the lower polarizer.

Figure 11-6 shows the tabulated data and extinction curves of a biaxial mineral. Note that the equatorial curve is not a great circle; this establishes that the mineral is biaxial rather than uniaxial. Another characteristic of biaxial extinction curves is that two of the principal vibration directions (Y and either X or Z) fall on the equatorial extinction curve, 90° apart. The third principal vibration direction (either X or Z) falls on the polar extinction curve.

The three principal vibration directions can be found graphically by a trial-and-error approach (Fig. 11-7). Rotate the tracing until a point on the polar extinction curve overlies the spindle's equatorial plane on the stereonet below (the single NS great circle on the stereonet). Then examine the great circle that is 90° away from this point. If the great circle intersects two points on the equatorial extinction curve, check to see whether these two points are 90° apart. Normally they are not; this means that the trial location was unsuccessful. Rotate the tracing until another point on the polar extinction curve overlies the equatorial plane of the stereonet. Again check the great circle 90° away from this point to determine whether it intersects the equatorial extinction curve at two points that are 90° apart. Continue until two such points are found; these two points, with the original point on the polar curve, are the principal vibration directions. This procedure takes only a few minutes after the routine has been learned.

One unfortunate aspect of the above procedure is that there are two solutions possible, one of which is false. Fortunately, the false solution is easily eliminated. Note in Figure 11-6 that the

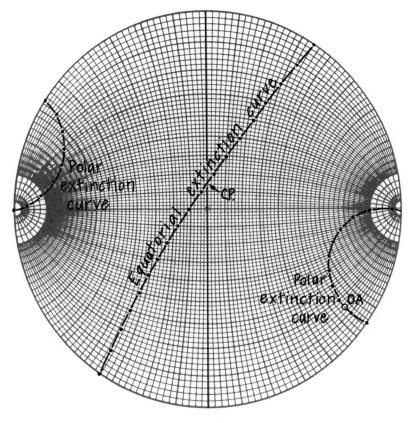

Figure 11-5.
Tabulated data and extinction curves for a *uniaxial* mineral (quartz). One of the extinction curves (either E_s or E_s^{**}) always passes through the plotted position of the spindle axis (at $E = 0°$); this is called the polar extinction curve. The second curve always crosses the spindle's equatorial plane (heavy line) at E_s or $E_s^{**} = 90°$. The crossover point is labeled as C.P.; the curve is known as the equatorial extinction curve. The equatorial extinction curve is a great circle for all uniaxial materials. This can be verified by rotating the tracing until this curve corresponds to an underlying great circle. The optic axis (OA) is the pole of this great circle (after Bloss, 1981, Fig. 2-5).

Mineral __Quartz__

Extinction Data ($\lambda = 589$)
$M_R = 3°$ $n_{Liq} = 1.544$

S	M_S	E_S	E_S^{**}
0	59°	56°	146°
10	60.5	57.5	147.5
20	63	60	150
30	66.5	63.5	153.5
40	70.8	67.8	157.8
50	76	73	163
60	82.8	79.8	169.8
70	89	86	176
80	96	93	3
90	103	100	10
100	109	106	16
110	115	112	22
120	119	116	26
130	122.7	119.7	29.7
140	125	122	32
150	127	124	34
160	127.4	124.4	34.4
170	127.4	124.4	34.4
180	127	124	34

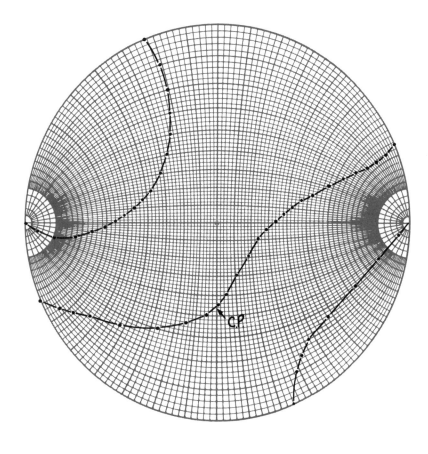

Figure 11-6.
The equatorial extinction curve of a *biaxial* mineral is not a great circle. Two of the principal vibration directions (Y and either X or Z) are located on the equatorial extinction curve. The remaining principal vibration direction (either X or Z) must lie on the polar extinction curve. These three directions must be 90° apart from each other (after Bloss, 1981, Fig. 3-2).

Mineral **Orthopyroxene**
Extinction Data ($\lambda = $)
$M_R = 4°$ $n_{Liq} = 1.690$

S	M_S	E_S	E_S^{**}
0	27	23	113
10	25	21	111
20	25	21	111
30	26	22	112
40	28	24	114
50	31	27	117
60	36	32	122
70	42	38	128
75	47	43	133
80	51	47	137
85	56	52	142
90	60	56	146
95	65	61	151
100	70	66	156
110	76	72	162
120	83	79	169
130	89	85	175
135	93	89	179
140	99	95	5
145	109	105	15
150	121	117	27
155	136	132	42
160	147	143	53
165	152	148	58
170	156	152	62
180	161	157	67

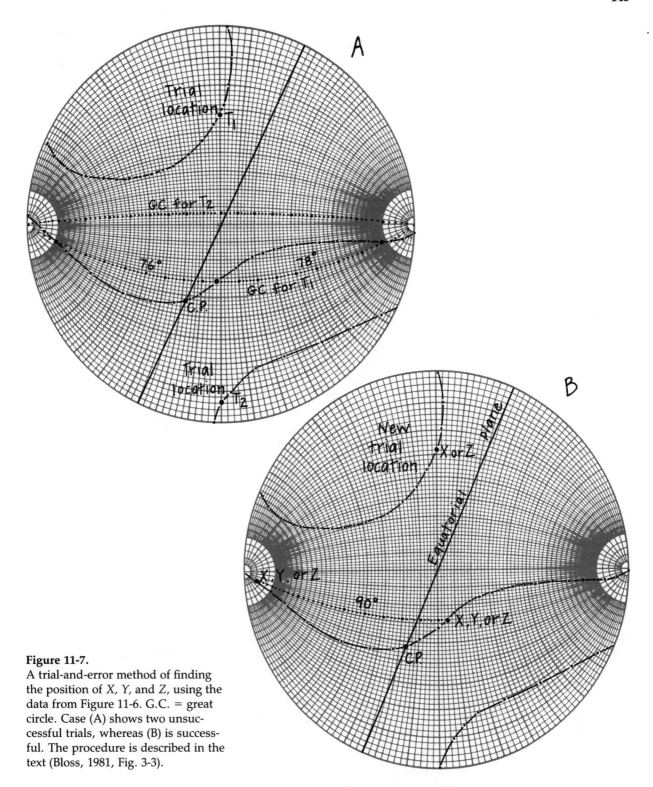

Figure 11-7.
A trial-and-error method of finding the position of X, Y, and Z, using the data from Figure 11-6. G.C. = great circle. Case (A) shows two unsuccessful trials, whereas (B) is successful. The procedure is described in the text (Bloss, 1981, Fig. 3-3).

equatorial extinction curve crosses the spindle's equatorial plane (labeled C.P. for the crossing point). If a false solution has been reached by the procedure just described, one of the three points found will be the point C.P. Therefore, before beginning the determinative procedure, clearly indicate on the tracing (Fig. 11-6) the point C.P. This point should not be taken as a possible location for X, Y, or Z. On the other hand, if an alternate set of locations for X, Y, or Z cannot be found, this means that one of the principal vibration directions has been set up perpendicular to the spindle axis, and the C.P. location is a correct solution.

After the three principal vibration directions have been located, their character (X, Y, or Z) can be determined either with interference figures or accessory plates (after appropriate rotations). Index measurements can be performed with the Becke test by rotating the desired vibration direction into the horizontal plane (after determining its S value) and then rotating the microscope stage until the vibration direction is parallel to the lower polarizer.

The angle $2V$ can be determined in a variety of ways, including direct measurement using interference figures, or from precisely determined values of α, β, or γ (using the Mertie chart, Fig. 8-7). It can also be determined, without knowing refractive indices, from direct measurements from the extinction curves, using the equation of Garaycochea and Wittke (1964, p. 185):

$$\tan^2 V_z = \frac{\cos \theta_z (\cos E_z - \cos \theta_z \cos E)}{\cos \theta_x (\cos E_x - \cos \theta_x \cos E)}$$

The approach is to pick any random point on an extinction curve that is neither coincident with the spindle axis nor located on an EW great circle that coincides with X, Y, or Z. Determine on the tracing the S and E values for the randomly chosen point. The term θ_z is the angular distance from the chosen point to the optical direction Z (whose coordinates are S_z and E_z). The term θ_x is the angular distance from the chosen point to the optical direction X (whose coordinates are S_x and E_x). After substituting the appropriate values into the equation, obtain the angle V_z, which is the angle between one optic axis and the optical direction Z. From this the $2V$ can be obtained. This calculation should be performed for a number of randomly chosen points and averaged to obtain the $2V$.

Instead of using the above approach, the data used for obtaining the extinction curves can be subjected to computer analysis using the program EXCALIBR, developed by Bloss (1981). The program will compute positions of X, Y, Z, and $2V$ with considerably greater precision than the graphical approach described above. Copies of the program (on nine-track tapes) can be purchased by writing to EXCALIBR, Department of Geological Sciences, Virginia Polytechnical Institute and State University, Blacksburg, VA 24061. Spindle stages can be purchased from Technical Enterprises, Inc., 2008 Carrol Street, Blacksburg, VA 24060.

THE UNIVERSAL STAGE

In contrast to the spindle stage, which is designed for the examination of single mineral grains, the universal stage is intended mainly for work with thin sections. Rotation of a thin section about a variety of axes permits measurements to be made between various planar or linear features and optical directions, both within and among mineral grains.

Description

Universal stages (Fig. 11-8) are mounted on the microscope stage; commonly a research-level microscope is required for this. Universal stages with three, four, or five rotational axes are available; four axes are sufficient for most applications.

A thin section is mounted in the center of the universal stage above a circular glass plate that fits in the central opening. A drop of glycerine is placed between the two glass surfaces in order to prevent abrasion during movement of the thin

Figure 11-8.
A modern four-axis universal stage (courtesy of E. Leitz, Inc., J. Hinsch).

section. In order to prevent light loss by reflection during tilting of the universal stage, glass hemispheres are attached below the central glass plate and above the thin section. Glycerine is applied to the glass-to-glass contacts. The attachment of the upper hemisphere, although snug, permits the thin section to be moved about on the plate.

In the initial or horizontal rest position, the rotation axes (from center outward), are oriented as follows:

A1   Inner vertical (IV)

A2   NS horizontal

A3   Outer vertical (OV)

A4   Outer EW (as distinguished from the inner EW of the five-axis stage)

In addition to rotations of the universal stage, the microscope stage (designated as M or A5) can also be rotated. The amounts of rotation of the various axes are read on adjacent scales or arcs.

The angle of inclination of the thin section can be determined directly from the scale readings. However, the ray of light through the sample may not be vertical because of refraction; refraction is encouraged at high angles of inclination, when the indices of refraction of the sample and the

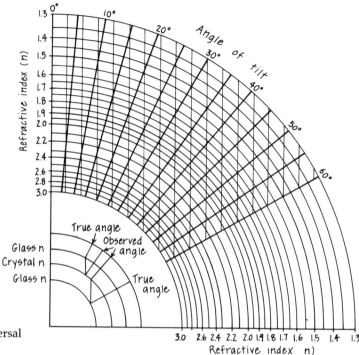

Figure 11-9.
Federow chart for correcting refraction in universal stage measurements.

mounting hemispheres are significantly different. Refraction effects can be corrected by the use of either Snell's law directly or the graphical solution of Snell's law (Fig. 11-9) as devised by E. V. Federow, the designer of the universal stage.

Special (telephoto) objectives with long working distances are necessary when using the universal stage, because of the presence of the upper glass hemisphere. These objectives contain an iris diaphram to control the effective aperture within the lens.

Universal stages are used with either orthoscopic or conoscopic illumination. For orthoscopic illumination, diaphragms are partially closed (to exclude convergent rays), and the larger hemispheres are used with low- or medium-power objectives. For conoscopic illumination, small hemispheres are used with high-power objectives (N.A. about 0.6) and a special condenser lens.

Observations

One of the common procedures is the *determination of plagioclase feldspar composition* by the Michel–Lévy method. This method measures the maximum extinction angle between α' and the (010) albite twin composition plane, and utilizes a determinative curve such as given in Volume 2 in the description of feldspars. After finding a plagioclase grain with albite twins, the IV (A1) axis is rotated until the composition planes are NS. The NS (A2) axis is then rotated until the composition planes are vertical; this is indicated by sharply bounded edges and a consistency of interference color in adjacent individuals. The extinction angle is measured between the low index direction α' and (010) by rotation of the microscope stage (A5). This angle is measured for a variety of settings of the outer EW (A4) axis until a maximum value is obtained. The plagioclase composition is determined by use of either a high-temperature (volcanic) or low-temperature (plutonic) extinction-angle curve. Plagioclase compositions can also be obtained by a variety of other methods involving angular relations between cleavage planes, twin planes, and optical orientation; the techniques are described in detail by Slemmons (1962).

The universal stage permits the *orientation of the indicatrix* of both uniaxial and biaxial grains to be determined orthoscopically or conoscopically. If a large number of grain orientations are measured on the same thin section, the data can be plotted on an equal-area projection, contoured, and analyzed statistically. This technique is commonly limited to uniaxial materials in deformed metamorphic rocks, and has furnished a great deal of the subject matter of petrofabrics (structural petrology). For more on the method, see Donn and Shimer (1958), Fairbairn (1949), and Turner and Weiss (1963).

After the optical orientation of a biaxial mineral has been obtained, it is often possible to determine the *optic angle* ($2V$) by appropriate manipulation. This can be done with either orthoscopic or conoscopic illumination. A knowledge of the $2V$ of an olivine is enough to determine its composition. For both orthorhombic and monoclinic pyroxenes, a knowledge of $2V$ in conjunction with a specific index determination furnishes a fair estimation of the composition.

The universal stage has been used to determine the precise orientation of cleavage and twin surfaces relative to the principal optical parameters. In addition, it provides a simple method of distinguishing rhombohedral carbonates in thin section (Emmons, 1943).

ADDITIONAL READINGS

Bloss, F. D., and D. Reiss. 1973. Computer determination of $2V$ and indicatrix orientation from extinction data. *American Mineralogist 58*, 1052–1061.

Bloss, F. D. 1981. *The Spindle Stage: Principles and Practice.* Cambridge, U.K.: Cambridge University Press, 340 pp.

Emmons, R. C. 1943. *The Universal Stage.* Geological Society of America Memoir 8, 205 pp.

Fairbairn, H. W., and T. Podolsky. 1951. Note on precision and accuracy of optic angle determination with the universal stage. *American Mineralogist 36*, 823–832.

Flinn, D. 1973. Two flow-charts of orthoscopic U-stage techniques. *Mineralogical Magazine 39*, 368–370.

Noble, D. C. 1968. Optic angle determined conoscopically on the spindle stage: II. Selected rotation method. *American Mineralogist 53*, 278–282.

Wilcox, R. E. 1959. Use of the spindle stage for determination of principal indices of refraction of crystal fragments. *American Mineralogist 44*, 1272–1293.

Bibliography

Bloss, F. D. 1961. *An Introduction to the Methods of Optical Crystallography.* New York: Holt, Rinehart and Winston, 294 pp.

Bloss, F. D. 1978. The spindle stage: A turning point for optical crystallography. *American Mineralogist 63,* 433–477.

Bloss, F. D. 1981. *The Spindle Stage: Principles and Practice.* Cambridge, U.K.: Cambridge University Press, 340 pp.

Carozzi, A., 1960. *Microscopic Sedimentary Petrography.* New York: Wiley, 485 pp.

Cayeux, L. (Carozzi, A. V., transl.) 1970. *Carbonate Rocks.* Riverside, N.J.: Hafner, 472 pp.

Deer, W. A., R. A. Howie, and J. Zussman. *Rock-Forming Minerals.* New York: John Wiley & Sons.
    Vol. 1, 1962: *Ortho- and Ring Silicates,* 333 pp.
    Vol. 1A, 1982: *Orthosilicates,* 919 pp.
    Vol. 2, 1963: *Chain Silicates,* 379 pp.
    Vol. 2A, 1978: *Single-Chain Silicates,* 668 pp.
    Vol. 3, 1962: *Sheet Silicates,* 270 pp.
    Vol. 4, 1963: *Framework Silicates,* 435 pp.
    Vol. 5, 1962: *Non-Silicates,* 371 pp.

Deer, W. A., R. A. Howie, and J. Zussman. 1966. *An Introduction to the Rock-Forming Minerals.* New York: John Wiley & Sons, 528 pp.

Dietrich, R. V., and B. J. Skinner. 1979. *Rocks and Rock Minerals.* New York: John Wiley & Sons, 319 pp.

Donn, W. L., and J. A. Shimer, 1958. *Graphic Methods in Structural Geology.* New York: Appleton-Century-Crofts, 180 pp.

Ehlers, E. G. 1980. Use of cleavage as an aid in the optical determination of minerals. *Journal of Geological Education 28,* 176–185.

Ehlers, E. G., and H. Blatt. 1982. *Petrology: Igneous, Sedimentary, and Metamorphic.* San Francisco: W. H. Freeman, 732 pp.

Emmons, R. C. 1943. *The Universal Stage.* Geological Society of America Memoir 8, 205 pp.

Fairbairn, H. W. 1949. *Structural Petrology of Deformed Rocks.* Reading, Mass.: Addison-Wesley, 143 pp.

Fleischer, M., R. E. Wilcox, and J. J. Matzko. 1984. *Microscopic Determination of the Nonopaque Minerals.* U.S. Geological Survey Bulletin 1627, 453 pp.

Garaycochea, I., and O. Wittke. 1964. Determination of the optic angle $2V$ from the extinction curve of a single crystal mounted on a spindle stage. *Acta Crystallographica 17,* 183–189.

Hartshorne, N. H., and A. Stuart. 1960. *Crystals and the Polarising Microscope,* 3d ed. London: Edward Arnold, 473 pp.

Hartshorne, N. H., and A. Stuart. 1969. *Practical Optical Crystallography,* 2d ed. New York: American Elsevier, 326 pp.

Heinrich, E. W. 1965. *Microscopic Identification of Minerals.* New York: McGraw-Hill, 414 pp.

Johannsen, A. 1918. *Manual of Petrographic Methods,* 2d ed. New York: McGraw-Hill, 649 pp.

Kamb, W. B. 1958. Isogyres in interference figures. *American Mineralogist 43,* 1029–1067.

Kerr, P. F. 1977. *Optical Mineralogy,* 4th ed. New York: McGraw-Hill, 492 pp.

Larsen, E. S., and H. Berman. 1934. *The Microscopic Determination of the Nonopaque Minerals.* U.S. Geological Survey Bulletin 848, 266 pp.

Laskowski, T. E., and D. M. Scotford. 1980. Rapid determination of olivine compositions in thin section using dispersion staining methodology. *American Mineralogist 64,* 401–403.

Laskowski, T. E., D. M. Scotford, and D. E. Laskowski. 1979. Measurement of refractive index in thin section using dispersion staining and oil immersion techniques. *American Mineralogist 64,* 440–445.

Mertie, J. B., Jr. 1942. Nomograms of optic angle formulae. *American Mineralogist 27,* 538–551.

Milner, H. B. 1952. *Sedimentary Petrography.* London: Thomas Murby, 666 pp.

Nassau, K. 1980. The causes of color. *Scientific American 230,* no. 10, 124–154.

Palache, C., H. Berman, and C. Frondel. *Dana's System of Mineralogy,* 7th ed. New York: John Wiley & Sons.
    Vol. I. 1944. *Elements, Sulfides, Sulfosalts, Oxides,* 834 pp.
    Vol. II. 1951. *Halides, Nitrates, Borates, Carbonates, Sulfates, Phosphates, Arsenates, Tungstates, Molybdates, etc.,* 1124 pp.
    Vol. III. 1962. *Silica Minerals,* 334 pp.

Parslow, G. R. 1977. A note on $2V$ estimation (simple technique for student use). *Journal of Geological Educators 25,* 17–18.

Phillips, W. R., and D. T. Griffin. 1981. *Optical Mineralogy, the Nonopaque Minerals.* San Francisco: W. H. Freeman, 677 pp.

Ramdohr, P. 1969. *The Ore Minerals and their Intergrowths.* New York: Pergamon Press, 1174 pp.

Rosenfeld, J. L. 1950. Determination of all principal indices of refraction on difficultly oriented minerals by direct measurement. *American Mineralogist 35,* 902–905.

Shelley, D. 1985. *Optical Mineralogy,* 2d ed. New York: Elsevier, 321 pp.

Slemmons, D. B. 1962. Determination of volcanic and plutonic plagioclases using a three- or four-axis universal stage. *Geological Society of America Special Paper 69,* 64 pp.

Troeger, W. E. 1979. *Optical Determination of the Rock-Forming Minerals,* 4th ed. (In English, by H. U. Bambauer, F. Taborszky, and H. D. Trochim). Stuttgart: Schweizerbart, 188 pp.

Turner, F. J., and L. E. Weiss. 1963. *Structural Analysis of Metamorphic Tectonites.* New York: McGraw-Hill, 545 pp.

Wahlstrom, E. E. 1979. *Optical Crystallography,* 5th ed. New York: John Wiley & Sons, 488 pp.

Wilcox, R. E. 1959. Use of the spindle stage for the determination of principal indices of refraction of crystal fragments. *American Mineralogist 44,* 1272–1293.

Wilcox, R. E. 1984. Refractive index determination using central focal masking technique with dispersion colors. *American Mineralogist 68,* 1226–1236.

Williams, H., F. J. Turner, and C. M. Gilbert. 1982. *Petrography.* San Francisco: W. H. Freeman, 626 pp.

Winchell, A. N. 1937. *Elements of Optical Mineralogy. Part I. Principles and Methods.* New York: John Wiley & Sons, 263 pp.

Winchell, A. N., and H. Winchell. 1951. *Elements of Optical Mineralogy. Part II. Descriptions of Minerals.* New York: John Wiley & Sons, 551 pp.

Winchell, A. N., and H. Winchell. 1964. *The Microscopical Characteristics of Artificial Inorganic Solid Substances: Optical Properties of Artificial Minerals.* New York: Academic Press, 403 pp.

Index

Definitions are indicated by italicized page numbers

Abbe refractometer, 19
Absorption formula, *56*
   determination of, 56–57, 133–134
Accessories, optical
   determination of optic sign
     biaxial figures, 108–111
     uniaxial figures, 69–74
   determination of vibration directions, 54–57
   first-order red plate, *53*–56
   gypsum plate, *53*
   mica plate, *54*–56
   quartz wedge, *53*–56
Acute bisectrix (Bxa), *83*–85
Analyzer, *3*
Angle (*see also* Optic angle, and Units and symbols)
   of incidence, *17*
   of reflection, *17*
   of refraction, *17*
Angular aperture, *92*–93
Anisotropic substances, *10*
Attenuation, *13*

Ballard, S. S., 15
Battey, M. H., 57
Becke, F., 21, 74
Becke, Line, 21–25
   associated colors, 23–24
   effect of dispersion, 23–24
   use of, 36–39
Berry, L. G., 57
Bertrand lens, *3*, 58–59
Biaxial substances, *26*, *81*
   identification, 126–137
   indicatrix, 82–84
   interference colors, 98–101
   interference figures, 91–113
   optic angle (2V), *83*–85, 93, 97, 103–104
   optic axes, *83*–85
   optic sign, *84*–85
   principal indices, *81*
   principal optical directions, *81*
   principal sections, *81*
   vibration directions, 85–90
Biot-Fresnel Rule, *94*–97
Biotite, absorption of, 33
Birefringence, *28*
   determination of, 78, 135
Bisectrix, acute and obtuse, *83*–84
Block diagrams, 127–133
Bloss, F. D., 5, 25, 125, 149
Bouma, B. J., 15
Brewster, D., 9
Brewster's Law, *9*

Calcite, double refraction, 31–32
Celestite, identification of, 127–128
Centering of objectives, 3
Chamot, E. M., 137
Circular sections of indicatrices, 33, 82–84
Circularly polarized light, *15*, 49–53
Cleavage
   basal pinacoid, 36
   hexagonal, 41–44
   monoclinic, 123–124, 128–130
   orthorhombic, 117–118, 127–128
   quality, 40–41
   rhombohedral, 35–36
   tetragonal, 41–44
   triclinic, 123–125
   type of extinction and, 42
   uniaxial grain orientation and, 40–44
   use in identification, 127–133, 135
Clinozoisite, identification of, 128–129
Colemanite, identification of, 129–130
Color
   mixing, 8
   origin by absorption and reflection, 10–11
   origin by electronic displacement, 11–12
   wavelength, 7
Compensation by accessory devices, 55
Complementary colors, 7–8
Condensing lens, 2, 58–59, 148
Conoscopic illumination, 58, 91
Converging lens, 2, 58–59, 148
Critical angle, *18*
Crossed nicols, *34*

Dana, E. S., 57
Delly, J. G., 137
Density of immersion liquids, 76–77
Detent spindle stage, 138–146

Detent spindle stage (*continued*)
  determination of optic angle, 146
  extinction curves, 141–146
  interference figures, 139–140
  orthoscopic observations, 140–146
  setting up, 138–139
Diamond pressure cell, 138
Diaphragm
  aperture, 2
  field, 2
Differential absorption of light, 9
  determination of, 56–57, 133–134
Dipole, atomic, 11
Dispersion
  coefficient of, 16
  curves, 17
    Hartmann equation, 17
  monoclinic, 119–123
    crossed, 119–120
    inclined, 120–121
    parallel, 122–123
  of index of refraction, 16, 116–125
  optical, 16
  optical orientation of, 119–125
  orthorhombic, 116–117
  staining, 25
  triclinic, 123–125
Ditchburn, R. W., 15
Double refraction, 31–32

Electromagnetic spectrum, 7
Elliptically polarized light, 15, 49–53
Elongation direction, 128–131, 135
Emmons, R. C., 25
Epsilon (ε), 28
  determination of, 38–39
Epsilon-prime (ε′), 28
  variability of, 28–29
Extinction
  cause of, 35
    inclined, 129–130
    parallel, 42–44, 118
    symmetrical, 42–43, 118
  use of, 35–39
Extinction angles, 118–120, 129–130, 135–136
Extraordinary ray (E), 30, 56
Eyepiece, 3

Fairbairn, H. W., 149
Federow correction diagram, 148

Federow, E. V., 148
Field lens, 3
First-order red plate, 53
  biaxial sign determination, 108–111
  determination of vibration directions, 54–59
  uniaxial sign determination, 68–74
Fisher, D. J., 25
Flash figure (*see* Optic normal interference figure)
Fletcher, L., 74
Flinn, D., 149
Fluorescence, 11
Ford, W. E., 57
Fraunhofer lines, 16
Frequency of light, 8

Gates, R. N., 25
Ghatak, A. K., 15
Grain mounts (preparation of), 18–19, 75–76

Hallimond, A. F., 5, 39
Hardness, 75
Hartmann equation, 17
Heating stage, 138
Hermann-Mauguin symmetry symbols, 114
Hutchison, C. S., 137
Huygens, C., 30
Huygens construction, 30–31

Immersion liquids, 19
  density of, 76–77
Inclined optic axis interference figure, 62–65
Index of refraction (*n*), 16
  accuracy of measurement, 138
  determination of, 18–25, 36–39
  determination with detent spindle stage, 138, 142, 146
  dispersion, 16–17
  variation in solids, 79–80
Indicatrix (optical)
  biaxial, 82–84
  circular sections, 33, 82–84
  uniaxial, 32–33
Infrared microscopy, 137
Interference color chart, 52–53, Plate I
  orders of, 53, Plate I
  use of, 51–55, 78
Interference colors, 35, 51–53

anomalous, 79
chart of, Plate I
in biaxial grains and figures, 98–101
in uniaxial grains and figures, 61–67, 78
use of, 53–56
Interference figures, Plate II
biaxial, 91–113
  acute bisectrix (Bxa), 91–100
  Biot-Fresnel Rule, 94–97
  determination of vibration directions, 94–100
  distinguishing among figures, 101–103
  estimation of optic angle (2V), 85–87, 93, 97, 103–104
  isogyre movement, 96–97
  obtuse bisectrix (Bxo), 91, 100–101
  off-centered types, 104–107
  optic axis (O.A.), 103–104
  optic normal (O.N.), 91, 101
  optic sign determination, 108–111
  symmetry of, 104–107, 117, 124–125
formation of, 58–59, 91–101
uniaxial, 58–73
  determination of anisotropic character, 68
  inclined optic axis, 62–65
  optic axis (O.A.), 59–62
  optic normal (O.N.), 65–67
  optic sign determination, 68–74
  relation to crystal orientation, 68–69
Interference microscope, 138
Interference of light
  monochromatic
    full wave retardation, 45–48
    half wave retardation, 48–49
    intermediate retardation, 49–51
    quarter wave retardation, 49–51
  polychromatic, 51–53
Isochromatic curves, 61
Isochrome, 61
Isogyres, 61–67, 94–98
Isotropic materials, 10, 13, 16
  distinguished from anisotropic materials, 35
  identification of, 25

Jones, N. W., 5, 25

Kamb, B., 101–103
Kennedy, G. C., 137
Korda, E. J., 125

Light (*see also* Dispersion and Polarization)
   color, 7–8, 10–11
   combination of, 7–8
   electromagnetic spectrum, 6–8
   energy of, 8
   frequency, *8*
   interference of, 13–15, 45–53
   monochromatic, *7*
   polychromatic, *7*
   sources, 7
   transmission, 10–15
   units of measurement, 6
   velocity, 8
   velocity change, 13–14
   visible, *6*
   wavelengths, 6–7
Longhurst, R. S., 15
Louisnathan, S. J., 125
Loupekine, I. S., 39
Lower polarizer, 2
   determination of privileged direction, 33
   standardized orientation, 33

Magnification, 3–4
Malus, E., 8–9
Mason, B., 57
Mason, C. W., 137
McCrone, W. C., 137
Melatope (*see also* Dispersion), *61*
   location in biaxial figures, 93–98
Mertie chart, 85–87
Mertie, J. B., 85–87
Mica plate, *55*
   determination of vibration directions, 54–59
Michel-Lévy chart, *52–53*, Plate I
   use of, 51–55, 78, 135
Michel-Lévy method of plagioclase determination, 148–149
Microscopes
   centering of objectives, 3
   magnification, 3–4
   maintenance, 4
   manufacturers of, 1
   parts of, 2–3
   uses of, 1
Miller indices, 41

Morse, S. A., 125
Mueller, C. G., 15

n (*see* Index of refraction)
Nicol polarizers, 2
   crossed nicols, 34
   uncrossed nicols, 34
Nicol, W., 2
Noble, D. C., 149
Nosepiece, 3
Nullification, 13–14
Numerical aperture, 3, *92–93*
   relation to optic angle (2V), 93, 97

Objective lens, 2–3
   centering of, 3
   numerical aperture, 92
   oil-immersion, 92
Oblique illumination method, 24
Obtuse bisectrix (Bxo), 83–85
Ocular, 3
Olivine, color origin, 10
Omega (ω), *28*
Optic angle (2V), *83–85*
   calculation of, 85
   determination with detent spindle stage, 146
   determination with universal stage, 149
   estimation in figures, 93, 97, 103–104
   Mertie chart, 85–87
Optic (axial) plane, *91*
Optic axes (*see also* Dispersion)
   biaxial, 83–85
   uniaxial, 26
Optic axis interference figures
   biaxial, 103–104, *110–111*
   uniaxial, 59–62
Optic normal interference figures
   biaxial, *101*, 110
   uniaxial, 65–67
Optic plane, *91*
   monoclinic, 119–123
   orthorhombic, 115
   triclinic, 123–125
Optic sign
   biaxial materials, *84–85*
      determination of, 108–111
   uniaxial materials, *28*
      determination of, 38, 68–74
Optical orientation
   hexagonal system, 29–33
   monoclinic system, 119, 128–130

orthorhombic system, 114–117, 127–128
   tetragonal system, 29–33
   triclinic system, 123–125
Ordinary ray (O), *30*, 56
Orthoscopic illumination, 58

Parslow, G. R., 104
Parting, *40*
Peck, A. B., 113
Permissible (privileged) directions, *28*
Phase difference, *48*
Phemister, T. C., 74
Phillips, W. R., 39
Photomicrography, 4–5
Photons, 6
Phyllosilicate, polarization of, 13
Planck's constant, 8
Pleochroism, *56*
   determination of pleochroic formula, 56–57, 133–134
Podolsky, T., 149
Polarization, 8
   atomic, 11–13
   circularly polarized light, *15*, 49–53
   doubly polarized light, *13*, 26–33, 35
   elliptically polarized light, *15*, 49–53
   origin of, 8–10
   plane-polarized light, *8*, 45–49
   rotation of, 48–49
   vibration direction, 8–9
   vibration plane, 8–9
Polarizer, 2–3
Polaroid, 2, *9*
Polished sections, 25, 75
Primary colors, 8
Principal plane
   biaxial, *81–82*
   uniaxial, *27–29*
Privileged (permissible) directions, *28*

Quadrant, notation and determination of, 63–65
Quartz wedge, *53*
   biaxial sign determination, 108–111
   determination of vibration directions, 54–57
   uniaxial sign determination, 68–74

Ray velocity surfaces (uniaxial), 29–31
Reflection, 10–11, 17
Refraction
  angle of, *17*
  critical angle of, *18*
Refractometer, 19
Reiss, D., 149
Relief, 19, *20*, *21*, 77
Retardation, 44–53
  addition and subtraction of, 69–74, 108–111
  biaxial figures, 98–99
  uniaxial figures, 61
Ruby, color origin, 11
Rudolf, M., 15

Saylor, C. P., 25
Schaeffer, H. F., 5
Shurcliff, W. A., 15
Sign (*see* Optic sign)
Skiodrome, *95–96*, 100
Slawson, C. B., 113
Snell, W., 17
Snell's Law
  calculation of 2E, 92
  correction of refraction, 149
  derivation, 17–18
  deviation from, 30–31, 33
Sodium vapor lamp, 50
Spindle stage (*see* Detent spindle stage)
Stage, 2
  universal (*see* Universal stage)
Stereographic projections, 95, 140–146
Stoiber, R. E., 125
Stout, J. H., 15
Superposition of light waves, 13–15

Thin-sections, *1*, 24–25, 75
  advantages of, 77
  determination of thickness, 77–78
  dispersion staining, 25
  mounting media, 24, 77
Tobi, A. C., 113
Tourmaline, differential absorption of, 9–10
Twinning, *79*, 137, 148–149

Ultraviolet microscopy, 138
Uncrossed nicols, *34*
Uniaxial substances, 26
  determination of indices, 36–39
  identification of, 75–80
  interference figures, 58–73
Units and symbols
  α (alpha index), *81*
  α' (alpha-prime index), *82*
  A.A. (angular aperture), 92
  Bxa (acute bisectrix), *83–85*
  Bxo (obtuse bisectrix), *83–85*
  β (beta index), *81*
  ε (epsilon index), *28*
  ε' (epsilon-prime index), *28*
  F.W.D. (free-working distance), *92*
  γ (gamma index), *81*
  γ' (gamma-prime index), *82*
  λ (lambda), *6*
  μ (mu), 6–7, 92
  n (index of refraction), *16*
  N.A. (numerical aperture), 92–93
  ω (omega index), *28*
  θ (theta angle), *28–29*
  2E (observed optic angle), 92–93
  2V (optic angle), *83–84*
Universal stage, 146–149
  description, 146–148
  determination of carbonate species, 149
  determination of optic angle, 149
  Federow chart, 148
  Michel-Lévy method of plagioclase determination, 148–149
  orientation of grains, cleavages and twin surfaces, 149
Upper polarizer, 3

Velocity of light, 8
Vibration
  direction, *8–9*
  plane, *8–9*
Vibration directions, 8–9
  biaxial interference figures, 94–100
  determination of, 33–39, 54–57
  uniaxial interference figures, 59–68

Wave front, *30*
Wave normal, *30*
Wavelength, 6–7
  distribution, *7*
  units of measurement, 6–7
Waves
  electromagnetic, 6
  interference of, 13–15, 45–53
  length, 6–7
  reflection and refraction of, 8–9
  velocity change, 13–14
  vibration direction, 8–9
  vibration plane, 8–9
Willard, R. J., 113
Winchell, A. N., 57
Wright, F. E., 74, 113

Zoltai, T., 15